Edmund Hillary

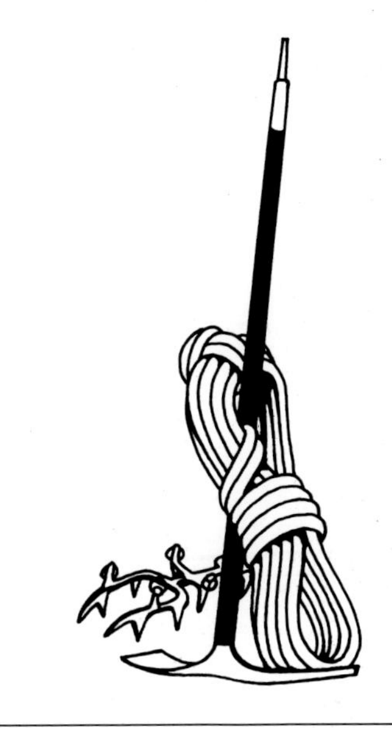

THE LIFE OF A LEGEND

ALSO BY PAT BOOTH

Fiction:
Dear Chevvy
Long Night Among the Stars
Footsteps in the Sea
Sprint from the Bell
Sons of the Sword

Non-Fiction:
The ABC of Injustice
The All Blacks' Book for Boys
The Bert Sutcliffe Book of Cricket
The Boot – Don Clarke's Story
The Fate of Arthur Thomas – Trial by Ambush
The Hunter and the Hill
The Islanders
The Mr Asia File
Jeff Wilson – The Natural
Graham Dingle – Dangerous Journeys

Deadline – an autobiography

Edmund Hillary

THE LIFE OF A LEGEND

Pat Booth

Hodder Moa Beckett

ISBN 1-86958-640-9

First published in cased edition 1993
Reprinted in 1994

Reprinted in limp edition 1997, 1998
by Hodder Moa Beckett Publishers Limited
[a member of the Hodder Headline Group]
4 Whetu Place, Mairangi Bay, Auckland, New Zealand

Typeset by Typocrafters Ltd, Auckland
Back cover image painted by Edward Halliday, 1955

Printed in Singapore by Kyodo Printing Co.

CONTENTS

FOREWORD

I believe that the loyalty and determination engendered by men like Shackleton . . . stemmed from their own great qualities as leaders and one such of this select group, I hold, is Sir John Hunt; another, Sir Edmund Hillary. There are those, it is true, who do not share this view of Hillary but I notice that they are rarely men who knew him well and are more often those who seem driven by an unreasonable and compelling jealousy to castigate any public figure or popular hero. Perhaps this is not surprising, as the very strength of his nature makes him enemies as well as friends. Fortuitous circumstance, some say, placed him on the summit of Everest — 'it might have been anyone'. Those who have experienced his formidable determination and turbulent energy doubt it. Like George Lowe, they feel that he was the right man, possibly the only man, in the right place at the right time.

Edmund Percival Hillary is a complex person with a depth of character which few people discern at a casual meeting. There is nothing flamboyant about him, absolutely no showmanship and, at least on the surface, he is no text-book hero. Indeed, he can be scornful of the more romantic conceptions of mountaineering. All the same, it is probable that in their profound sincerity, his own conceptions of exploration and climbing mountains are the most romantic of all. To this man, adventure itself, and the ability of man by physical and moral training to conquer the erstwhile impossible, are the basic ingredients of life.

He is not a demonstrative person — credit and praise from him are never misplaced and are infinitely more valuable than the same sentiments expressed by the more expressive or less dedicated men. Family life is very important to him — he is devoted to his own — and one senses in him a deep moral conviction, essentially old-fashioned and sometimes even prim. He belies an apparent vanity by a disarming facility for self-criticism. He may not know all his own weaknesses but is quite unresentful of criticism . . .

He hates the gushing, the fawning — and that constant lack of anonymity which must accompany such people as himself — but accepts it now as unavoidable. He is a man of balanced judgment and great friendliness, yet he can appear surprisingly thoughtless, even careless of the feelings of others . . . He has been described as 'a virtuoso, deeply and constantly embroiled in his art'. Like Peter Paul Rubens, 'he is a truly golden character'.

Peter Mulgrew

(from *No Place for Men*)

PREFACE

When the twentieth century lists its achievers, the name of Sir Edmund Hillary will be among them, ranked with the greatest adventurers of this and many other centuries. But he will be honoured for much more than simply courage, tenacity and success.

He will be seen, too, as a man who honoured his debts, who carried off fame with modesty, who bore disaster with dignity and without rancour. A man prepared not only to challenge nature with courage and determination, but also to take up its case in an era in which Earth came under seemingly constant threat from its own inhabitants. He pushed back the boundaries of the known world, to the heights of Everest and to the depths of the polar snow, and also widened mankind's horizons both by what he achieved and by what he said and wrote.

Feted by world leaders, deeply admired by his peers, Hillary always carried with him the recognisable characteristics of the young, unknown New Zealand beekeeper who was the first man to lever himself to the top of the world's highest mountain.

So much changed in the forty years that followed. Hundreds followed Hillary and the Sherpa Tenzing to that peak, sometimes dozens in one day. When Hillary's son Peter made the climb a generation later, he called his father in Auckland, New Zealand, by satellite phone from the peak. That was the measure of the change.

But Ed Hillary remained much the same man who had made that first joint climb, who in his own simple idiom 'knocked the bastard off' and yet who, inevitably, had felt his life change from those minutes on the highest point of the world. Like astronauts who have looked back on their world from space, or walked the surface of the moon, Ed Hillary was all at once the same and yet never the same again. It was as if he had glimpsed more than the vast snow profiles of the Himalayas and had seen beyond to man's real place on the planet as caretaker in trust.

Through his unfailing support and nurturing of the Sherpa population, building schools and hospitals, repaying not only his own debt but also those of other people to other communities, he set an example of caring to the world. Through his worldwide travels in support of environmental causes, Hillary has come to epitomise the modern adventurer who not only travels but also defends, who not only sees for himself but also brings the message home to hundreds of millions more.

He is, in the very real meaning of the words, a man of the world.

A career built upon courage and challenge, the outpourings of the spirit that took him literally where no one had been before, has become a lifetime of compassion for humanity, its environment and the animal forms with which it shares this planet. These are now his life just as the literal peaks of achievement once were.

Forty years after those moments shared by a shy Sherpa and a young beekeeper handpicked by Fate to be the trailblazers, the time is right to celebrate a great life.

No book of this type would be possible without the co-operation — sometimes anonymous — of a number of people. Those I can thank by name include George Lowe, Jim McFarlane, Michael Gill, Graeme Dingle, Rex Hillary, Robyn Mulgrew and Peter Hillary . . . and Sir Edmund Hillary himself, for the time he has given over the years to me and to other writers.

My thanks, too, to the staff of both the Auckland Central Library and News Media Library, Auckland, for their help with research, to the photographic editors at the *New Zealand Herald*, News Media, Wellington Newspapers, *The Press*, the *Christchurch Star* and the Royal Geographical Society, for assistance in illustrating the book, and to Severn House Publishers and Reed Publishers for their permission to quote from the writing of the late Sherpa Tenzing and Peter Mulgrew.

Pat Booth

CHAPTER ONE

'High Dream in the Sky'

Many times I think of that morning at Camp Nine. We have spent the night there, Hillary and I, in our tent at almost 28,000 feet which is the highest that men have ever slept. It has been a cold night. Hillary's boots are frozen and we are almost frozen too.

But now in the grey light, when we creep from the tent, there is almost no wind. The sky is clear and still. And that is good.

We look up. For weeks, for months, that is all we have done. Look up. And there it is — the top of Everest. Only it is different now — so near, so close, only a little more than a thousand feet above us. It is no longer just a dream, a high dream in the sky, but a real and solid thing, a thing of rock and snow. This time, with God's help, we will climb on to the end.

Then I look down. All the rest of the world is under us. To the west Nuptse . . . straight down the ridge, two thousand feet down, is the South Col where our nearest friends wait — Sahibs Lowe and Gregory and the young Sherpa Ang Nyima, who yesterday helped us up to Camp Nine. Below that is the white wall of Lhotse, four thousand feet more, and at its bottom, the Western Cwm, where the rest of our friends wait at the advance base camp. Below the Cwm is the Icefall, below the Icefall the Khumbu Glacier.

I see that Hillary is looking too and I point. Below the glacier, 16,000 feet down, you can just see in the grey light the old monastery at Thyangboche.

To Hillary, perhaps, it does not mean much. To a man from the West it is only a far, strange place in a far, strange country. But for me it is home. Beyond Thyangboche are the valleys and the villages of Solo Khumbu and there I was born and grew up. On the tall hill-sides above them, I climbed as a boy, tending my father's yaks.

Home is close now. I can almost stretch out my hand and touch it. But if it is close, it is also far. Much further than 16,000 feet. As we strap on our oxygen tanks I think back to the boy, so close and so far, who had never heard of oxygen, but yet looked at this mountain and dreamed.

Then we turn around, Hillary and I. We begin to climb.

It is many miles and many years that have brought me here.

Sherpa Tenzing's account of one morning in 1953
(from *Man of Everest*)

At 6.30 a.m., we crawled out of the tent and stood on our little ledge. Already the upper part of the mountain was bathed in sunlight. It looked warm and inviting but our ledge was dark and cold. We lifted our oxygen tanks on to our backs and slowly connected up the tubes to our face-masks. My 30 lb load seemed to crush me downwards and stifled all enthusiasm but when I turned on the oxygen and breathed it deeply, the burden seemed to lighten and the old urge to get to grips with the mountain came back. We strapped on our crampons and tied on our nylon rope, grasped our ice-axes and were ready to go.

Edmund Hillary's account
(from *High Adventure*)

Two men on a mountain. Below them the monastery. Above them the sky. The rock, the snow and the ice in a final peak, white surfaces marked with their boots, scarred by their ice-axes, soft snow with the shape of their gloves forever frozen.

Around them the familiar, unconquered walls and the steep traverses, the ridges on which others had stumbled and never risen again, remembered names now in a grave of ice. The gleaming bulk of Makalu and Kanchenjunga and Cho Oyu, the massed peaks of the Himalayas stretching away to the lost horizon beyond.

Two men tired in their triumph, not seeing it that way — for they were climbers rather than poets — but caught in a moment of history. Like Columbus and the Americas, like Cortez and the Pacific, like Stanley and Livingstone, Scott and Amundsen, and Armstrong who would later leave his footprints on the moon.

History could wait — not for long as it turned out — because for those moments they were simply themselves. For the last time, but they were not to know that.

Simply two men. On a peak called Everest, 11.30 a.m., 29 May 1953.

The mass of Everest dominates one of the expedition camps. First called simply Peak XV, it was named Everest in 1865 after the surveyor-general of India, Sir George Everest.
ROYAL GEOGRAPHICAL SOCIETY

On 2 June I had been given my assignment — and a very small part in a very big event — breaking the news to the Hillary family.

The address was written on the carbon copy of the Reuters message but I didn't trust myself to find the place. I was still a country boy at heart and Auckland seemed a wasteland of suburbs. So I followed what I knew to be the Remuera tramline slowly up the Meadowbank hill.

The house was as quiet as the garden around it when I knocked and I could hear the sound echo inside.

High on that peak, the two men were at first almost unable to grasp it all. After the physical strains and the mental tensions of the hours before, suddenly it was over. They had achieved what had challenged and defeated so many. There was no doubting that. The spread of ice and snow below was proof.

For both there was a surge of satisfaction unmatched for them before or since. Almost solemnly, in that typically understated New Zealand/British style, Ed Hillary extended his heavily gloved hand. They shook hands vigorously.

It was not enough. Suddenly, Tenzing allowed the emotion in both of them to run free. He threw his arms around his companion's shoulders and they stood high above the world thumping each other on the back. Ed Hillary has never forgotten that broad smile across Tenzing's ice-caked face.

The face that came around the half-opened door was hesitant and inquiring, for no other reason, I suspect, than that she wanted to resist unwanted salesmen or collectors. Gertrude Hillary was always polite. She was a slim, groomed woman with inquiring, intelligent eyes and a half-smile.

The tension went quickly and the beginnings of a smile became a look first of astonishment and then absolute delight when I introduced myself and hurried into my news. Beside me, a flashbulb flared before the doorway was suddenly emptied. I heard her calling up the stairs.

'Perce, come quickly. Edmund is at the top!'

On the peak, they became conscious time was running out for them. They needed to leave enough oxygen for the descent. Before they had left their final camp, Tenzing had rigged up flags on the handle of his ice-axe: the Union Jack, the UN flag, Nepal and India. Ed Hillary hadn't bothered to take a New Zealand flag with him; nationalism was not an issue.

Nor was personal fame. Ed Hillary's photographs of Tenzing and flags unfurled on the peak are the only record of their joint achievement. He hadn't thought to give Tenzing instructions on using a camera and somehow that was neither the time nor the place.

Edmund Hillary and Sherpa Tenzing pause on the way to Camp Nine as they move towards their final assault. Although at more than 27,000 feet, Hillary has taken off his oxygen mask for a few minutes. Behind his left shoulder is the peak of Makalu.
ROYAL GEOGRAPHICAL SOCIETY

Gertrude Hillary hears of son Edmund's feat at her front door from young *Auckland Star* reporter Pat Booth, 2 June 1953.
NEWS MEDIA

There was joy but no emotion in the Hillary sitting room in Auckland. Percy had taken a minute or so to appear. I almost had the feeling that he had stopped to put on a tie for the occasion.

He ordered tea for the visitors while he read and re-read the Reuters cable: 'Everest, the highest mountain in the world, has been conquered. The men who reached the peak were Edmund P. Hillary of Auckland and Tenzing, the leading Sherpa porter. The Times of London announced that a dispatch had been received from Colonel John Hunt, leader of the British expedition, saying the summit was reached last Friday. Colonel Hunt added: "And all is well."'

When Gertrude emerged from the kitchen with the cups and biscuits, she paused — a mother alert to the welfare of her son — to read that line again. 'And all is well.'

'Thank God,' she said, and she meant it.

The moment of triumph — Tenzing on the top of Everest. Edmund Hillary had not arranged for Tenzing to photograph him!
ROYAL GEOGRAPHICAL SOCIETY

God was not forgotten on Everest either. With the minutes ticking away, Tenzing scratched a hole in the snow for small offerings of sweets for the Gods of Chomolungma, whom Buddhists like him believed lived on the peak. Earlier, Tenzing had prayed to them: 'God of my father and mother be good to me now — today.' The gods had looked after them well in those last hours and would continue their protection on the descent.

Ed Hillary placed beside the offering a small white crucifix which a British priest had sent to expedition leader John Hunt asking that it be left on the world's highest point. Also left were a coloured pencil Tenzing had promised he would leave for his daughter Nima, and a small, stuffed-toy black cat John Hunt had given Hillary as a good luck charm.

Both men carried back pebbles from the peak. Forty years later, I put a reverent finger on one, set in classic Nepalese workmanship and hanging in the Auckland lounge of Rex, Ed's brother, beside a photo of the peak it came from.

Looking back later, I realised that Gertrude Hillary wanted to get to the phone. Instead, she tried to steady the obvious excitement that made her voice tremble at times, and kept up a flow of good-hostess small talk.

Percy returned from a few minutes rummaging in a cupboard. 'Edmund is a good man under pressure.' He had the evidence in his hand, photographs from his son's album showing a rescue party on New Zealand's Mt La Perouse five years before and a figure in a makeshift frame being winched up an ice face. Wrinkled clippings told the story.

A young medical student, Ruth Adams, had fallen 200 feet while attempting Mt La Perouse in February 1948, breaking an arm and suffering other injuries and concussion. Ed Hillary was one of a party which turned back from a climb to carry the injured Ruth clear. He dug an ice cave while others went for help.

For three nights, Hillary and another climber clung there, keeping up Ruth's spirits, trying to deaden the pain in her arm and to keep her warm. Ed Hillary stripped off some of his heavy mountain clothing to help Ruth fight the cold. After two days, a plane dropped supplies.

When a sixteen-man rescue party reached them, they were forced to carry her on a stretcher the 1000 feet up to the peak to lower her down the ice and snow of the western side. The descent took four more gruelling days.

One of the small pebbles of Everest rock that Hillary brought back from the summit.
NEWS MEDIA

The boots that took him to the top — the size 12, three-pound boots Hillary wore, now an exhibit in Canterbury Museum. The old-style pint milk bottle is there to provide a point of scale.
THE PRESS

Five years later, married to a doctor, with a daughter and living in London, Ruth was one of those who hailed the Everest climb as fitting reward for a job well done.

Percy Hillary stared at the photos as his wife Gertrude piled the cups and eased her way towards the phone. 'Edmund,' he said, 'always carries his weight.'

On Everest, Ed Hillary linked past with the present and the future. He spent a little time looking about for any sign of the legendary Mallory.

George Mallory, the mystic of mountaineering, friend of poets Rupert Brooke and Robert Graves, was the man whose classic reply to an obvious question became, like him, part of climbing history. While he was an undergraduate at Oxford's Merton College, he practised climbing by scaling the college chapel, crossing its roof and descending the opposite wall. In the 1920s, he began his love affair with Everest. During the 1922 expedition, he was a survivor when an avalanche swept seventeen climbers before it, killing seven. Undeterred, Mallory returned two years later. Asked why he wanted to climb Everest, he replied: 'Because it's there.'

One June day in 1924, Mallory and fellow climber Andrew Irvine disappeared into the mists and cloud, were last sighted climbing at around 27,000 feet and were never seen again. How did they die, and where; on the way to the summit or even returning from it? That day, thirty-nine years later, the often blizzard-swept peak gave no hint of their fate.

Everest still holds its dead, now and then revealing them in their sleeping bags, hunched in the remains of tents, a woman caught in climbing position on one ice face, another cross-legged, mouth open apparently screaming into the cold that killed her.

The peak Mallory called Pumori (Tibetan for little one), to remind him of his daughter, still looks on. It was from these slopes that Everest veteran Eric Shipton and a young climber called Hillary first talked of climbing Everest from the south. But of Mallory there is nothing. Only the mountain whose call he could not resist.

Percy Hillary was still shuffling the photographs, some of them murky snaps of high country huts, groups struggling across rocky streams, parties moving through snow and across glaciers.

Would son Edmund find it hard to come back after all that had happened? Percy shook his head. 'He's a very hard worker with the bees. I'm sure he'll be glad to get back to it.'

A supply party moving up the lower part of the Western Cwm towards Camp Four during the assault on Everest.
ROYAL GEOGRAPHICAL SOCIETY

The honey factory in Papakura to which Percy Hillary believed his son would return.
NEWS MEDIA

On the high peak, Ed Hillary paused to look out towards the bulk of Makalu, the world's fifth highest peak at 27,790 feet, seeking and finding a likely route to the summit. The possible assault path he saw in his mind's eye was later used by the French in their successful ascent. Beyond, the two men could see for more than a hundred miles, but they could not see into the future: the life-and-death struggles on Makalu; a son's bids to follow his father to the highest point they now shared. For that day, there was only Everest.

They sat in the snow and ate a bar of mintcake. It seemed like a victory banquet. As they stared about them, they occasionally glanced at each other, savouring the inexpressible satisfaction of it all. They had come so far, not just on this climb but on the whole journey of their life.

Ed Hillary, then aged thirty-three, was a beekeeper and amateur mountain climber from New Zealand, a place Tenzing could not even visualise.

The Sherpa Tenzing was a Tiger — one of those tough, determined and skilled mountain people who had been the literal backbone of this and every past Everest expedition. The Tiger title was a symbol of his skills, given by the Himalayan Club. Tenzing, one of thirteen children, aged thirty-nine, was on his seventh visit to Everest. He first joined the 1936 reconnaissance, had climbed the east peak of Nanda Devi in 1951 and a year later was with the Swiss expedition.

For him, 29 May 1953 was doubly memorable. Exactly one year before, to the very day, he had stumbled down the mountain, exhausted, with the Swiss guide Raymond Lambert. They had been beaten back from a point on the south-east ridge only about 1000 feet from the summit on which he now stood. His great companion Lambert was very much with him in those moments. On the climb to the top, he and Hillary had found oxygen bottles abandoned on that attempt. There too was the twisted skeleton of the tent in which the two men had sheltered one year before, now only frayed fragments of orange material whipped by the wind.

Tenzing had told his new companion only hours before of the misery of those hours. 'We had no sleeping bags . . . all we ate was a little cheese washed down with snow melted over a candle . . . we lay as close as possible together rubbing and slapping each other to keep our blood flowing.'

Once out of their tent in even worse weather next day, Lambert and Tenzing climbed 650 vertical feet in five hours. They could go no further.

> I believe in God. I believe that in man's hardest moments He sometimes tells them what to do and that He did it then for Lambert and me.
>
> We could have gone further. We could perhaps have gone to the top. But we could not have got down again. To go on would be to die . . . and we did not go on. We stopped and turned back.
>
> We had reached an altitude of about 28,250 — the nearest men had ever come to the top of Everest, the highest anyone had ever climbed in the world. But it was still not enough. We had given all we had and it was not enough.

A year later, the memories of Lambert and that failure added a further dimension to the day. Tenzing wore to the top a red scarf Lambert had worn a year before and then given to his Sherpa friend. Later, Tenzing sent it back to him as a reminder of their friendship and the glory which could so easily have been shared by them.

Other very practical mementoes went with him on the great climb — his old Swiss reindeer-leather boots, a woollen helmet given him by Earl Denman, an eccentric Canadian-born adventurer from Africa who, in 1947, made an attempt on Everest with only Tenzing and one other Sherpa in support, and socks his wife Ang Lahmu had knitted.

With him, too, went those earlier realisations of the perils of the great mountain: the deaths of seven Sherpas on Everest in 1922; how six others perished with four German climbers on Nanga Parbat, one gallant Sherpa choosing to die with the German leader high on the slopes rather than leave him and go to safety. Tenzing had also been close to the bizarre tragedy of Maurice Wilson. Wilson had gone secretly to Tibet in 1934 to attempt a solo climb. Tenzing was one of the 1935 expedition that discovered an old tent and the body of Wilson, bent over as if trying to take his boots off, on the North Col. They buried him where he died.

Neither Hillary nor Tenzing had any doubts about the risks. But they did have a confidence in their individual and joint skills borne out by the scale of their feat of that morning. Their attitudes reflected much of their backgrounds and their beliefs.

Hillary would say later:

> I'm afraid that [after childhood] I drifted away from religion. Although I am interested in philosophy as such and in many religions, I don't

The British flag flies high over the Everest expedition's second base camp.
ROYAL GEOGRAPHICAL SOCIETY

The man and the mountains. Edmund Hillary at 27,200 feet.
ROYAL GEOGRAPHICAL SOCIETY

Following pages: The 1953 Everest expedition's first base camp at Thyangboche, near the famous monastery.
ROYAL GEOGRAPHICAL SOCIETY

Hillary in the striped sunhat that became his trademark around the world. It was made by his sister June from children's clothing.
ROYAL GEOGRAPHICAL SOCIETY

George Lowe, always irrepressible, relays a message to lower camps from high on Everest, as Hillary laughingly approves.
ROYAL GEOGRAPHICAL SOCIETY

personally have any deep religious beliefs. I don't believe in God in the ordinary church sense of God. I believe that the world is so complex and so remarkable in many ways, the whole universe, that there must be some sort of intelligence behind it all.

But whether that intelligence is the slightest bit interested in some little person down here on earth, I have considerable doubts.

When I began mountaineering and was on a difficult climb or in a hazardous situation, potentially dangerous or likely to be avalanched off something, because of my background I sometimes had the feeling that maybe a little bit of appeal to God might help and I might safely get through.

But I quickly built up the attitude that if I was in trouble then that was my business. I had to get myself out of it. There was no use me getting down on my hands and knees and praying to God. It wouldn't do me the slightest bit of good.

The best of luck to people who are having problems and wish to pray to gain help. Maybe it's a slightly arrogant approach I've had but I've always felt 'Ed, you're in this problem. You've got to get yourself out of it.'

For his part, Tenzing, who had begun the day with prayer, prayed again.

It was such a sight as I had never seen before and would never see again — wild, wonderful and terrible. But terror was not what I felt. I loved the mountains too well for that. I loved Everest too well.

At that great moment for which I had waited all my life, my mountain did not seem to me a lifeless thing of ice and rock, but warm and friendly and living . . .

. . . as I covered up the offerings, I said a silent prayer. And I gave my thanks. Seven times I had come to the mountain of my dream and on this, the seventh, with God's help, the dream had come true.

Thuji chey, Chomolungma. I am grateful.

Hillary's much more matter-of-fact account:

It was 11.30 a.m. My first sensation was one of relief — relief that the long grind was over; that the summit had been reached before our oxygen supplies had dropped to a critical level; and relief that in the end the mountain had been kind to us in having a pleasantly rounded cone for its summit instead of a fearsome and unapproachable cornice. But mixed with relief was a vague sense of astonishment that I should have been the lucky one to attain the ambition of so many brave and determined climbers . . .

After fifteen minutes, the two men adjusted their masks. Tenzing untied the flags from his ice-axe, spread them across the snow of the peak, and they began the descent.

New Zealander George Lowe's account of their arrival:

At 1 p.m., they appeared again on South Summit and began the descent of the steep loose snow slope . . . they stopped at Camp Ten at 2 p.m. . . . just after 4 p.m. I set out again to meet them . . . they were moving fairly rapidly — the only tiredness showed in their slightly stiff-legged walking as they cramponed the last bit . . . Ed unclipped his mask and grinned a tired greeting, sat on the ice and said in his matter-of-fact way: 'Well, we knocked the bastard off!'

The triumphal but tired return to the base camp after the Everest climb — Hillary and Tenzing (both in goggles) with Sir John Hunt (left), Tom Bourdillon and George Lowe (extreme right, carrying the rope).

ROYAL GEOGRAPHICAL SOCIETY

CHAPTER TWO

Peak Times

If an imaginative scriptwriter had dreamed it up, they couldn't have done it better. First the coded message. This was the apparently disappointing report of an assault repulsed sent by radio from *The Times* correspondent James Morris: 'Snow conditions bad stop advance base abandoned yesterday stop awaiting improvement stop All well.'

Back in London, a pre-arranged cipher told the real story, complete with names and date. Then, the five-ring circus bulletin revealing all on 2 June, to coincide with the coronation of the young Queen Elizabeth II, the news I, for my part, carried in a rental car to an address in Remuera, Auckland, New Zealand.

But even the most imaginative scriptwriter would have stopped short of the final twist a few years on. James Morris had a well-publicised sex change operation to become Jan Morris. This was the new byline the account of this subterfuge carried when he/she wrote it twenty-five years later. Also in his/her accounts of the expedition was a description of the Khumbu icefall on the approach to Everest: 'Like going up the kitchen stairs for three or four miles at a go, three steps at a time — and carrying the baby.'

The Morris personal headlines lay ahead; in the present was instant stardom for a man who might have been specially cast as the archetypal New Zealander. Ed Hillary, Papakura beekeeper, was six foot three, lean, craggy-faced, laconic, modest, good-humoured and perfectly cast for the role. Many media photographs of him in the days after the climb showed him wearing a striped sunhat his sister June had made from one of her children's playsuits. That was Ed. He looked the product of a copybook colonial background — and was.

Gertrude Hillary's family, the Clarks of Northern Wairoa, were

Yorkshire stock and farmers, a close family of twelve children who rallied around their widowed mother after father Clark was killed in his fifties by a kicking horse. Gertrude, among the youngest, trained as a teacher and later taught.

The Hillary family were also from a Yorkshire migrant last century, a watchmaker who married an Irish governess. Ida Hillary was a woman of character in a colonial community peopled by men and women with guts and drive. Her grandson Rex still treasures examples of her painting, detailed early bush scenes that hang on the walls of his home alongside Sherpa art and photos of high peaks. Much younger than her husband, Ida Hillary set a pattern of powerful, influential women who held their families together, who led but did not dominate, who widened the boundaries of their children's lives.

Ida also preserved history in a way which had particular significance in 1953. Gertrude Hillary unfurled for me Ida's carefully prepared family tree, in particular its documentation of the link with a nineteenth-century hero, Sir William Hillary, whose great-great-grandnephew climbed Everest.

William Hillary earned a baronetcy as a tribute to his leadership of a private army during the Napoleonic Wars. He spent thousands of pounds to arm and organise 1400 cavalrymen and infantry — the largest private army of its time in Britain — and then went to war himself to lead it.

On retirement to the Isle of Man, he helped save 300 men and women from shipwrecks. Once, with only his son to help him, he set out in a tiny boat to bring in seventeen men from a sinking ship. He ventured back out into another storm only weeks later with six broken ribs for another rescue. In 1823, he appealed to the nation for help in establishing a lifeboat and rescue service. He led by example. In 1830, he won the Shipwreck Institution's gold medal for his part in saving sixty-two survivors of another foundering ship.

When his poor health stopped him taking his place at the oars, he spent more of his fortune building refuges for shipwrecked victims near deadly shoals, and homes for seamen at major ports. And he was often on the beach to help survivors ashore when the lifeboats came in.

Ida's recording of his life reflected her pride. Gertrude's recall of Ida's life carried that theme forward another hundred years. Ida's son, Gertrude's husband, Edmund's father, inherited rather the erratic characteristics of his father, who, family legend has it, retired in his mid-sixties and went to bed for thirty years.

Percy was independent in action and thought — to an extreme. As a sixteen-year-old, he tried to enlist to go to the Boer War. When he finally made it to the Great War, he was badly wounded on Gallipoli. He was married in uniform after being invalided home.

Left: Ida Hillary, who passed on to famous grandson Edmund her strength, determination and independence.
REX HILLARY

Right: Percy and Gertrude Hillary, parents of a son to be proud of.
REX HILLARY

Later he was for a time a country newspaper editor at Tuakau, forty miles from Auckland. A man with an enthusiasm for reading, he built a bookrack to hang over the backs of the family's few cows as he milked them. He and Gertrude had three children, Edmund Percy, born on 20 July 1919, Rex, born sixteen months later, and older sister June. Percy's beekeeping hobby became his livelihood after, inevitably, he fell out with the newspaper's directors.

They lived what would be seen now as an austere and simple life, growing their own food and making honey from an increasingly large number of hives, caught up in a series of the health and spiritual beliefs common in those times.

Percy was a hard man. Sixty years later, Ed Hillary talked feelingly of his clashes with him, of fierce rows with a man who was a harsh disciplinarian at times, of beatings in the woodshed. A strict father was at odds with a son who acknowledges now that he was stubborn and playing out battles of wills.

'I was a very restless, dissatisfied, lonely child . . . I always had this feeling that life was restricted, that I wasn't seeing enough and doing enough and there must be more to it than that.

'He was a man of great energy, he was very restless. He tried to inflict very rigid views on his family. My teens were full of conflicts with him. When I look back on it, of course, I must have been an absolutely appalling child because I did argue about everything.

'But I never lost my respect for my father, because he was a tremendous worker and a doer of things — though he didn't always finish things he was a great starter and stimulator.

'I always admired the fact that even though I mightn't agree with what he said, I knew he would be working harder than I would and would be following through with his principles very strongly.'

Their mother was their inspiration. She particularly stressed the value of education and as a former teacher did much of it herself at home. Percy, like many fathers of the times, wanted children who could contribute to the family budget. Gertrude demanded the schooling he said they could not afford. Her will prevailed, but sometimes only just. Rex

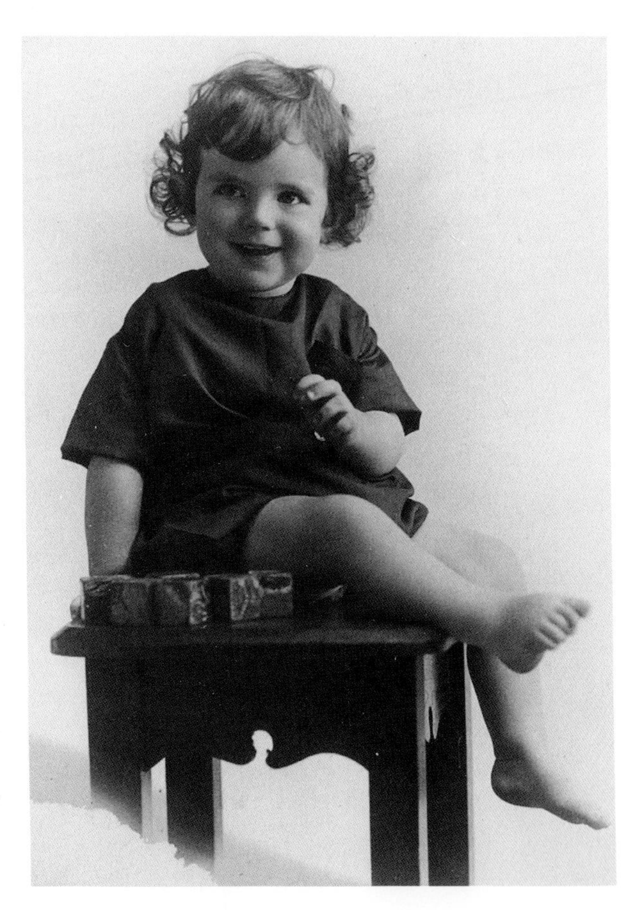

Ed Hillary as a curly-haired, chubby two-year-old — the lean look came much later.
REX HILLARY

actually left school at one stage and then returned after a year. He remembers a family environment where their father seemed to take it in turns to bear down on one of the children. One or the other was always out of favour.

Close in age, the two brothers were also close in many ways. They were for years at the same school, Tuakau primary. Urged on by his mother, Ed was — as he laughingly described himself more than sixty years later — 'the child genius of Tuakau'. He passed proficiency at eleven, two years ahead of the normal age. He told me once that his mother had 'pumped me full of rather useless information and the teachers at Tuakau liked teaching me because I could absorb information'.

Tuakau School.

REPORT on Edmund Hillary, Std. V.

for Term ending 22nd August, 1930.

ATTENDANCE ½ days out of 126.

Marks obtained in Examination at end of Term.

		OUT OF	
Reading and Recitation	14/64	100	G.
Spelling	21	25	V.G.
Writing	19	25	G.
Composition	80	100	V.G.
Grammar	45	50	Ex.
Arithmetic	97	100	Ex.
Drawing	31	50	
Geography	26	50	
History and Civics	33	50	G.
Conduct: V. Good.			
Industry: Excellent.			
Total ...	430	550	

Place in Class 1st.

Remarks Edmund thoroughly deserves his place in class. He has worked well all the term.

Class Teacher.

F. N. R. Downard. Head Teacher.

Signature of Parent or Guardian P. A. Hillary

Well done, Edmund!

Champtaloup & Edmiston, Stationers, Auckland

How Tuakau School rated a pupil who would make history.
NEWS MEDIA

A long line of Sherpas carrying supplies towards the sites of the high camps a month before the assault on Everest.
ROYAL GEOGRAPHICAL SOCIETY

Hillary adjusts Tenzing's oxygen equipment as the pair prepare to leave for the high-altitude camps.
ROYAL GEOGRAPHICAL SOCIETY

Again, Gertrude's will prevailed and the young Ed was sent to Auckland Grammar, forty miles away in the big city. Looking back, this decision probably played a major part in moulding him. His school day spanned eleven hours including up to four hours on the train. Inevitably, he was a loner. He describes his time at Grammar as very unhappy, particularly the early stages when as a small, much younger boy from a totally rural background he tried to foot it in a big city school with boys much older and much larger. He was shy and felt small and rather helpless.

His real pleasure was back at Tuakau, sharing with Rex the verandah bedroom, its open end curtained off against the weather. They played their own simple games with few toys, spent much of their time running around the countryside, making their own life — including the time that older sister June tried to teach them to dance, first essential in an adult social life. She gave up with them well short of Fred Astaire standard.

More than sixty years later, he recalls very precisely one incident, which he says affected him 'and still affects my whole attitude to life'.

'When I first went to Grammar, we all had to go along to the gymnasium to be assessed for sporting potential. This gymnastic instructor, whom I regard as one of the more unpleasant teachers I ever had anything to do with, looked at me when I stripped off.

'I clearly remember him staring at me with scorn and saying "What will they send me next?" He told me I had a bulging rib cage and my back wasn't straight, everything that was terrible about my physical appearance and set-up. I was mortified and this created in me an enormous sense of inferiority which, even although I developed into a rather large and robust person, I still retain.

'I think this incident built up in me — I was after all only 11 at the time — a determination that I would become competent in something.'

Later, in the same interview, he directly related that putdown in the gymnasium to the success on Everest. 'I felt it proved that I wasn't such a loser as that fellow back at Grammar had suggested.' He was still feeling the sting of it.

In fact, the eleven-year-old weakling matured very quickly. The heavy work Percy required of his sons around the hives helped the young Ed to grow quickly — six inches in his fourth-form year, five in the next. There grew too a fierce competitiveness, even with brother Rex in their beekeeping routines. Rex looks back:

'With the bees, we used to compete, one against the other. When we went out to look through the hives during the spring, bending over the hives all day, going from one site to another, we'd compete, at extracting honey from our 1300 hives.'

Ed remembers this trait. 'Even when we were walking, I had to be in front.' That desire to be first!

From another interview: 'There's no question that I had this sort of basic inferiority and I was constantly having to prove to myself that I wasn't as humdrum as I felt deeply in my heart I was. I still feel this applies. My life has been a constant effort to illustrate how a very mediocre person with very mediocre talents, which I have, can create quite a lot if they really drive themselves.'

The two brothers were growing in more than their bodies. They were also developing strength of character and conviction. The Hillary family would these days be regarded as 'New Age', heavily into unusual philosophic and physical beliefs, exponents of what was then known as 'Radiant Living', dabbling in theosophy and alternative religions.

Out of all this came a strong pacifist belief which saw first Ed and then Rex register as conscientious objectors to war service. Percy also applied to have Ed excluded as working in significant wartime work. Given a dispensation, Ed worked around the hives.

When Rex's call-up came, he wasn't so fortunate. The examining board felt one brother excused from service was enough. He was sent to a conscientious objectors' detention camp at Strathmore, between Rotorua and Taupo, for four years. The jingoistic community of that time was not an easy society in which to parade conscientious objection, but the two brothers did — Rex more publicly since he was marched off in what was regarded by 'right-thinking' people as public disgrace. He remained in detention until the end of the war.

Edmund later revised his thinking and served as a navigator in Catalina flying boats in the South Pacific. It was, he said later, the best holiday of his life; the heavy fighting had by then moved to the north. He was finally invalided home with burns after an explosion and fire in an air force launch after the Japanese surrender.

His appetite already well whetted by climbing experience both before and after air force service, the twenty-six-year-old Ed literally headed for the high country. These were formative, empowering years in which he learnt new skills and how best to make use of his physique. The rejected schoolboy had grown into a lanky, powerful man with an urge to go places and the strength to do just that.

Sometimes Rex went with him. There were excursions to Egmont, Ruapehu (which Ed had visited first with a grammar school party), the Kaikouras, the Southern Alps. Often, the expeditions were in winter when the bees were dormant. They were, but Ed certainly wasn't. Wherever there was height, you'd find a Hillary or two. Only when the pace got too hot, the peaks too high, when the demands Ed made on himself and his companions increased, did Rex retreat to what became his basic and vital role in the Hillary success — bringing in that 300 tons of honey to put the Hillary label on it.

Long before Everest, the solemn Hillary brothers, Edmund and Rex.
REX HILLARY

Leading Aircraftsman Ed Hillary in training at Delta Camp near Blenheim during war service.
NATIONAL LIBRARY OF NZ

The young Ed Hillary, photographed for the family album on two of the many climbing expeditions in New Zealand that ultimately led him to the top of Everest.
REX HILLARY

Harry Ayres, the man Hillary rates as New Zealand's finest climber and his mentor, studies the Tasman Glacier with his young pupil.
NEWS MEDIA

Ed had by this time made one of the two significant climbing contacts of his life, with legendary climber and guide Harry Ayres, whom Hillary described as 'the outstanding New Zealand climber'. It takes one to recognise one. It was Harry Ayres who led the party on Mt La Perouse that rescued Ruth Adams after her fall in 1948. So often, it was Ayres and Hillary on New Zealand's most challenging peaks.

Anything could lead to climbing. By 1949 his sister June was in London completing a Master's degree and due to marry an English doctor. Percy and Gertrude went off to the wedding. Summoned to drive them on a trip around the Continent, Ed dutifully obliged — then linked with New Zealanders Cecil Segedin and Bruce Morton on a climbing binge in the Alps that included five 10,000-foot peaks in five days.

Better was to come, and George Lowe was the source. The two had met in typically New Zealand style a few years earlier. They had been in the same bus in the Mt Cook area when they got to talking. George Lowe, reacting to Hillary's home being in Auckland, offhandedly mentioned that his father on their Hastings farm also kept bees as a sideline and got his queens from someone called Hillary in Papakura. 'That's us,' said Ed. 'My name's Ed Hillary. Small world, isn't it?'

Between them, they were to prove in the next few years just how much that small world was shrinking. They kept in touch through the climbing grapevine, shared a snowed-in hut in the Southern Alps on one climb when the party played draughts, using pieces of carrot and parsnip as pieces and a roughly drawn chequerboard which was, in fact, the back of a very outdated calendar.

On another trip, Ed Hillary, clearly much less conservation-minded than he was to become, stalked and killed a kea, a native New Zealand parrot, only to discover that, magnificent as they might seem in life when feathered, they were puny and inedible later.

It was about this time, perhaps walled in by snow or walking a glacier, the idea came that maybe they should try the Himalayas. From a chance remark, an idle dream, it became reality. A letter from George Lowe tracked Ed Hillary down in Switzerland suggesting he join a group planning to climb in Nepal. They were, at that stage and from the safe planning point of New Zealand, literally aiming high. There was idle talk of Kanchenjunga or even Everest.

In the end, they settled for less. Four climbers, Ed Hillary, George Lowe, Earle Riddiford and Ed Cotter, made up the first-ever New Zealand expedition to the Himalayas in 1951. They had already tried out themselves and their companion with tough climbs in the Southern Alps, enough to raise their hopes, enough for Ed Hillary to mark George Lowe as 'apart from Harry Ayres, who had no equal, the most competent climber I have been with'. It was an opinion that never wavered then or later. The four achieved their less extravagant ambitions, climbing their prime objective Mukut Parbat, 23,760 feet, and five other peaks of better than 20,000 feet.

This novice group were not going unnoticed. As they returned to their hill station base, destiny came calling again, just as it had before, by letter to Hillary in a Swiss chalet. A letter from the famed Eric Shipton invited two of them to join his exploration expedition to the Everest area — 'bring own food and supplies'.

Later, both Ed Hillary and George Lowe would tell their separate versions of the effect this letter had on the group — two invitations to be shared among four. Lowe described it as being like a Mills bomb thrown into their midst. Earle Riddiford, Lowe said, rationalised it immediately. He (Riddiford) would be one, the issue was who would be the second? In the end, the two were the one with the best cash reserves (Riddiford) and the strongest (Hillary).

It was a situation that could have tested most friendships, but after some heartache George Lowe and Ed Cotter gloomily waved their friends away on their journey to join Shipton and packed up for New Zealand. Neither really had much option; George Lowe had put his entire savings of £150 into the £1300 needed to fund the foursome, and needed to get back to his schoolteaching. While Everest was calling Ed and Earle, New Zealand bank managers were looking forward to the return of George and Ed Cotter.

They did not realise at the time, but Ed Hillary and George Lowe would yet both share an Everest experience. Better than that, it would be the definitive expedition to the Mother of the World.

The young New Zealander Hillary obviously quickly lived up to the good reports Eric Shipton had had of the group, and the two men — separated by geography, age and experience — formed a close link. They shared what was perhaps the most significant moment in Everest expedition history, the discovery of the southern route. Before 1951, all attempts on the great peak had been made from the north, as Nepal was at that time closed to foreigners. In any event, the southern approaches looked difficult, even impassable. Inspired by a glimpse of the mountain from the high slopes of 23,190-foot Pumori, both the Himalayan veteran Shipton and his younger teammate Hillary believed they had glimpsed a new route.

Two weeks later, Hillary and Shipton linked again with another reconnaissance of the area. Both tough, fit and competitive, they found the difference in their ages (Shipton was forty-four, Hillary thirty-two) was no handicap. They matched each other in a tough schedule. More importantly, they confirmed the apparent feasibility of a new approach to Everest, involving the Western Cwm (which Mallory had named years before, using the Welsh word for an enclosed valley in a setting a long way from The Valleys), the Khumbu icefall, the Lhotse Face, the South Col and on across the South Summit.

Both men believed they had found the answer to Everest. Determined to set up an assault by that route, Shipton took a party back the following year, 1952. The short-term objective was an attempt on Cho Oyu, 26,870 feet. In fact, the party was training for a full-scale attempt on Everest the following year. This time, Hillary and Lowe had no problems of acceptance. Both were named by Shipton, along with Earle Riddiford.

Cho Oyu survived. Hillary and Lowe were part of an assault team that attempted the peak; the New Zealanders were finally forced back at 22,500 feet by what Hillary later described as 'a real death trap of wall, ice cliffs . . . hanging over it great fractured fingers of ice which periodically avalanched and swept remorselessly down to the glacier 3000 feet below'. George Lowe had always remembered how Hillary said, on their first climb together in the Southern Alps of New Zealand, 'I don't think a climb is

really worthwhile unless you have been scared out of your wits at least twice.' Cho Oyu looked a prime contender for that definition.

A few weeks, and a couple of 20,000-plus peaks later, Hillary and Lowe were putting the Hillary 'scared out of your wits' theory to a further test in a first crossing of the Nup La pass which leads to the north face of Everest. This was a real challenge: a great icefall, a huge rock buttress, crevasses, unstable snow, hidden falls, actual and potential. But beyond it lay the inspiring north face of Everest and a feeling that, in making history with this first crossing, they were also touching the past as well. There were the last traces of the traditional first base camps of those attempts from the north side, bleak and still inhabited by the ghosts of men and hopes long gone — Mallory among them.

Meanwhile, on the southern approaches, the Swiss attempt by Tenzing and Lambert had been repulsed. They had ignored the path that Hillary and Shipton had surveyed and had instead attempted another route. The two groups, the Swiss downcast in failure and the British party stirred by their ambitions, shared a meal and talked . . . what else but Everest? Hillary came away moved by the courage of the Swiss and more determined than ever that the peak would be beaten.

Yet another attempt was made later that year, again with Lambert and Tenzing at the heart of it, this time attempting the Shipton-Hillary line, only to succumb to the intense cold as winter approached.

Everest was still, in the words of Mallory, 'there'. This time though, there would be no Shipton. Unexpectedly, unacceptably on first hearing by Ed Hillary, his respected leader and companion from those competitive excursions of a few months before, had been replaced by someone called John Hunt, a soldier, a colonel, with what Hillary later understated as 'a couple of Himalayan trips to his credit'.

With hindsight, Hillary later accepted that the Everest Committee had reason for the change. Hunt was an organiser and a strategist, a controller and a leader who could and did inspire men. Shipton, a magnificent and experienced climber, was a man for small groups, out of his depth with the sort of larger-scale enterprise the 1953 assault on Everest would become.

Nor was Hunt the apparent nobody Hillary had believed at first hearing. He was an army man through and through — Sword of Honour at Sandhurst, a wartime commander of the 11th Indian Infantry Brigade in Italy, a winner of the DSO. He was Army, yes, but that was no criticism. He was actually preparing for Nato manoeuvres in Germany when the invitation to succeed Shipton reached him.

The route of the first ascent of Everest — up the Khumbu Valley and Glacier (from lower right), through the Icefall, which turns sharply to the right, across the Western Cwm (partly hidden by cloud) to the Lhotse Face. Between Lhotse and Everest is the South Col.
NEWS MEDIA

Hillary's rather dismissive 'couple of Himalayan trips' reference did Hunt far from justice. He had been climbing since 1925 as a fifteen-year-old and had ten Alpine summer seasons in his diaries. In India between the wars, Hunt had taken part in three Himalayan expeditions and, like Wilfred Noyce (34), another of the party, had trained troops in snow and mountain warfare. He had used army postings to climb all around the world. He had been selected for an Everest party in the 1930s but had been rejected by a medical board.

As well as being a committed climber and excellent organiser, he was also a very good judge of men, their abilities and their potential. He named, for his Everest attempt, men like Dr Charles Evans (33), who had been in the Himalayas three times before and in three successive years — with Tilman on the Annapurna range to Kulu in 1951 and with Shipton on Cho Oyu — and would two years later climb Kanchenjunga. Tom Bourdillon (28) had been with Shipton on both the reconnaissance and Cho Oyu. Alfred Gregory (39) had also been on Cho Oyu.

And then, the two New Zealanders. This is how Hunt described Hillary (33): 'His testing in the Himalayas had shown that he would be a very strong contender, not only for Everest but for an eventual summit party. When I met Shipton I well remember his prophesying this — and how right he was. Quite exceptionally strong and abounding with restless energy, possessed of a thrusting mind which swept aside all unproved obstacles, Ed Hillary's personality had made its imprint on my mind . . .'; and Lowe (29): 'His New Zealand alpine experience dates from before that of Hillary, to whom he introduced some of the high standard climbs on those mountains. His ice technique, acquired like Hillary's from the exceptional opportunities offered by New Zealand mountains, is of a very high standard . . .'

Others in the party were: Charles Wylie (32), a serving officer in the Brigade of Gurkhas and a former prisoner of war of the Japanese; Michael Westmacott (27), 'a mountaineer of the first rank' according to Hunt's assessment, and an ex-president of the Oxford University Mountaineering Club; George Band (23), the baby of the party and vice-president of the Cambridge Mountaineering Club; Michael Ward (27), the doctor and a fine climber, who had suggested the reconnaissance of the south side of Everest two years before; physiologist Griffith Pugh; photographer Tom Stobart; and Tenzing, who was invited to join the climbing party several months after the first selection.

John Hunt's assessment of his team was obvious. They would not have been there had he not believed in their character and their skill.

But what was the team's view of their leader? Inevitably with the New Zealanders, some reaction to his military background hung on — briefly.

George Lowe recounted later how he and an equally sacrilegious

The 1953 Everest expedition stands behind its high altitude Sherpa team in a victory photograph after the successful assault: Tom Stobart, a Sherpa, Charles Evans, Charles Wylie, Ed Hillary, John Hunt, Tenzing, George Lowe, Michael Ward, Tom Bourdillon, George Band, Griffith Pugh, Alf Gregory and Wilfred Noyce.
ROYAL GEOGRAPHICAL SOCIETY

Hillary had rolled out of their tent and presented arms to Hunt with their ice-axes on one of the first days on the slopes. It went down like a brick. But John Hunt obviously shrugged it off and the two wild colonials quickly came to admire the leader with deep and lasting sincerity.

Only much later — presumably through the family information service after he married Hunt's daughter — did George Lowe discover that he and Ed Hillary had at one stage slipped off the team list. John Hunt thought their geography was against them, preventing them joining in pre-expedition training in Wales. Only a near-mutiny from other team members forced him to change his mind. Lowe bore no grudge.

This is how Lowe summed up Hunt after Everest:

> He was universally regarded (and so many people told me this) as the stiff, unbending military commander, a just man but formal and unemotional, the efficient army colonel.

> John Hunt was the idealist who could on occasions be brimful of temperamental feeling and sentiment.
>
> Far from being the clipped, orthodox soldier, he won the warmth of your heart as few commanding officers could hope to do, captivating his party with his own dreams.
>
> One among Hunt's ideals might have been but was never called trite. After the success of Everest he publicly pushed the well-worn platitude that it was 'team work' and 'team effort' on the mountain which had made the Everest dream come true. This was not only fact: it was an essential basis for reaching the summit.
>
> And yet, endlessly put forward to the world by a lesser spirit, the simple fact could have sounded not merely trite but tawdry. Not for an instant was it seen thus, for John Hunt had made Everest into a message, a goal for all human beings. Every man and woman could have their own little Everest . . .

This was the essence of the man who became Sir John (later Lord) Hunt of Everest and this was the spirit of that rather disparate group of men who gathered on the slopes of Everest in May 1953.

And Hillary? Well, he is not a man given to great praise or sentiment, even to his own family. But what he said has significance.

> I first met John Hunt in Kathmandu and despite my pre-conceived prejudice was immediately impressed . . . John had great energy and drive and expressed a complete conviction that our party could get to the top — an approach I always viewed with caution.
>
> He handled most of us with considerable skill — we were all individuals with a dislike of military procedures but we received no orders, only suggestions or requests, which were generally so soundly based that we were happy to agree and give our loyal cooperation. I learned to respect John even if I found it difficult at times to understand him . . .

The two assessments by Lowe and Hillary tell as much about the personalities of the authors as they do about the subject — Lowe outgoing and warm, Hillary reserved and often sparing in his praise.

Only once did the Hillary guard slip. In the last paragraph of his description of the climb in his contribution to Hunt's book *The Ascent of Everest*, Ed Hillary allowed himself the luxury of emotion in a few words untypical both in what they said and the way they said it: 'To see the unashamed joy spread over the tired, strained face of our gallant and determined leader was to me reward in itself.' Gallant, determined, John was both these things, and painstaking too. Perhaps never more so than in the weeks and days before the assault.

The two key decisions he faced involved who to use — the makeup of the assault teams — and which form of oxygen assistance they would use. The Hillary hope was, of course, that he would team with Lowe. Hunt did not take up that option, preferring, according to Hillary, to 'spread the vigorous New Zealand icemanship around the team'.

Evans and Bourdillon were named to make the first assault, aiming for the South Summit. If that was attained and more was possible, they would then strike out for the summit. They would use the closed-circuit oxygen system, in which the climbers inhaled a high concentration of oxygen from a breathing bag and exhaled through a soda lime canister that absorbed the expired carbon dioxide and allowed exhaled oxygen to return to the breathing bag.

This was a much more complex and potentially more fault-ridden system than the open-circuit method in which climbers inhaled air enriched by added oxygen and expired into the atmosphere. This was the system the second assault party of Hillary and Tenzing would use.

With the men and the methods selected, Hunt and the party set about establishing the high camps the attempts would require. In that process, this expedition set records undreamed of in previous attempts. Seven parties had attempted the north-east face between 1921 and 1938 and three more the south-east in 1951 and 1952.

In 1953, seven climbers reached heights well over 27,000 feet; four climbed the previously unbeaten South Summit. First to the South Summit, Evans and Bourdillon narrowly failed to go all the way on 26 May. After an heroic bid, hindered by fresh snow that made movement slow, they were beaten back, a few hundred feet and three days short of being the first men to the top.

In fact, Bourdillon had wanted to press on alone from a point above the South Summit but Evans convinced him that they had neither the strength in reserve nor the oxygen to make the return journey if they went further. Instead, they retreated down the mountain, emotionally and physically exhausted, falling often and at the point of collapse by the time they staggered into Camp Eight.

Two days later, Hillary and Tenzing positioned themselves for their attempt. Ahead of them, George Lowe had cut steps and carried gear with Alfred Gregory and the Sherpa Ang Nyima. When the party settled on a camp site, Lowe offered to stay on at that height to help the assault group down next day.

As Hillary tells it: 'With a lump in my throat I thumped him on the shoulder in appreciation and shook my head. A hearty handclasp with them all and Gregory led off wearily down the mountain — a tired but watchful George going down last.' Last man to go, first man to meet them, George Lowe epitomised so much about friendship and the teamwork that

John Hunt valued so much. When, next day, George Lowe walked out to meet Hillary and Tenzing on their return, Hillary's laconic 'we knocked the bastard off' was in part a re-statement of that team ethos — 'we' meaning not only the assault team but men like Lowe.

Lowe had spent eleven days above 23,000 feet directing the construction of the approach up the Lhotse Face, as long as any mountaineer in history, Hunt would say later. He told the *National Geographic* how Lowe had 'combated piercing cold, blizzards that obliterated overnight all the previous day's painstaking trailmaking, sickness that depleted his work party and the fearful deterioration wrought by altitude itself'.

Neither Lowe's high spirits nor his appetite, long legendary, waned at these heights. But he found sleeping difficult. Joined by Noyce one night, he took sleeping pills. Barely wakened by the next noon, he reeled along the trail and during a pause for food dozed off upright with a sardine hanging out of his mouth. He was a powerhouse on the upper slopes. *National Geographic* headlined this part of the account 'Lowe, hero of the Lhotse Face.' That was no mistake. Then Lowe, with the Sherpa Ang Nyima, climbed to exhaustion carrying supplies.

The teamwork did not stop there. Evans and Bourdillon, literally on their knees at times in their valiant failure — 'We were too dulled for disappointment, that came later' — brought back much practical detail about that South Summit where no man had ever been before, the last stretch of an approach to Everest that Mallory once talked of as 'one of the most awful and utterly forbidding scenes ever observed by man'.

They had seen it close up and, unlike Mallory, had returned to describe it.

Slow, careful progress across a crevasse on the Khumbu Icefall below Camp Two.
ROYAL GEOGRAPHICAL SOCIETY

CHAPTER THREE

Off-peak Politics

After the euphoria, after the tiredness, the drama and the tremendous sense of personal and shared achievement, came the aftermath. Significantly, it was George Lowe who recorded the reactions of both the men he met that May morning.

He quotes Hillary as being 'grumpy' over the knighthood, the fact that he was not asked whether he wanted it, that the New Zealand Government accepted on his behalf, that his first knowledge of it was a letter sent up the mountain to him by John Hunt, the envelope addressed to Sir Edmund Hillary, KBE. Hillary thought it was a joke. When he realised that it wasn't, he reacted.

'It should have been a great moment but instead I was aghast. It was a tremendous honour, of course, but I had never really approved of titles and couldn't imagine myself possessing one. I had a vivid picture of walking down the main street of Papakura dressed in my torn and dirty overalls and thought I'd have to get a new pair.' He said he went to bed that night feeling 'miserable rather than pleased'.

He, it turns out, was not the only person to have regrets. Lowe again reports Tenzing as being overwhelmed by the sequels to Everest and saying: 'I wish I'd never climbed it.'

Not that this reaction was obvious — or for that matter present — in the first days after the climb. The teamwork on which John Hunt (by then Sir John) had laid so much successful stress held together on the mountain.

The pressures came and the beginnings of problems were obvious when the party began what should have been a journey of triumph from the mountains. It *was* such a journey for Tenzing, but at the expense of John Hunt and Edmund Hillary who were with him, three of the team. Unfortunately, the nationalistic Nepalese did not see it as teamwork.

The summit edge of Everest from the South Summit, the view Hillary and Tenzing saw before they began their historic climb.
ROYAL GEOGRAPHICAL SOCIETY

A cup of tea for the Everest conquerors after their return from the top of the world.
ROYAL GEOGRAPHICAL SOCIETY

Ed Hillary, with sister June and brother Rex on a family outing to Waiheke Island long before Everest made him famous.
REX HILLARY

With Everest climbed, and Ed now Sir Edmund, the three meet again in Britain at the time of his investiture.
THE PRESS

They had their private moments — Tenzing walked to his mother's village to tell her his news. All her life she had believed there was a golden sparrow and a turquoise lion with a golden mane on the summit. She was delighted he was now free of his obsession but puzzled about her lost beliefs.

Then came their first days as world celebrities, and with the fame the first hints of problems. The fiercely nationalistic Indian and Nepalese Press wanted to represent Everest as a national triumph for either or both countries. (Tenzing was born in Nepal and had lived extensively in India.) They wanted literally above all else for him to have been the first man on the summit. Hunt, for his part, wanted to preserve the team philosophy. Things went from bad to worse and the aftermath lasted a very long time.

Nearly forty years later, Edmund Hillary sat in the sitting room of his Remuera home, rich in mementoes of Nepal, the Sherpas and Everest. A carved table at his sprawling feet was a gift from the head lama of Thyangboche monastery, whose monks had years before prayed each morning for two men's safe return. On the walls were rugs, and everywhere were memories of another place.

Amongst the memories was vivid recall of that journey from the mountain in 1953 and the rampant nationalism that swirled around them, the huge crowds and the absolute adulation of Tenzing. 'Long live Tenzing' — Hillary could look after himself. Above them were banners extolling the success of Tenzing, including some showing the Sherpa on top of a high mountain waving the Nepalese flag. From him dangled a long rope supporting a sprawling figure some distance below. No guesses who that figure was intended to be!

Hillary had no obvious rancour — 'at first I found it quite funny' — but it was obvious from his manner that the humour of it had worn thin very quickly.

He talked about one reception in particular where Tenzing was asked for a few hesitant words, cheered to the echo. He was followed by Sir John Hunt, who was listened to quietly but without great reaction. As Hunt walked to the backrow seating he shared with Hillary, a tug of war developed over the next speaker. One group of the organisers suggested it be Hillary. Another was adamant it should not be. When one finally settled it, he simply added to the atmosphere. He was sure, he said to the crowd, that they would like to hear from 'the second man to reach the top of Everest'.

All those years later, Ed Hillary could still remember the absolute silence as he walked to the dais. All he could remember was the sound of his footsteps there and back. He still talked of 'those people pouring out hate at me', and their fear that he would somehow publicly challenge this nonsensical myth about Tenzing.

In an effort to defuse the situation, both Hillary and Tenzing signed a statement that they had reached the top 'almost together', which neatly sidestepped the fact that — to be absolutely, even pedantically, correct — Hillary had made the final step with Tenzing fractions of a second later.

What did it really matter? Well, it clearly mattered to the Nepalese and Indians. Their reaction was fuelled by the world Press, and finally even an irritable Sir John Hunt didn't help. Pressed for an assessment of Tenzing, he told the media that the Sherpa was 'a competent climber within the limits of his experience'. It was a half-truth from a man clearly at the end of his tether. It was so unlike him and it was so clearly different from his assessment of Tenzing in the first weeks on the mountain and which he repeated a few months later in his personal account of the expedition: 'Tenzing is not only the foremost climber of his race but (also) a mountaineer of world standing.' That's what he should have said in those strange June days, but he didn't, and the fire of controversy continued to throw out heat and smoke.

Tenzing's naive, simple background was a major factor. The media could and did manipulate him. But he was also nobody's man. When Sir John Hunt tried what we now call 'damage control', seeking to contain Tenzing within the contracts that gave *The Times* sole rights to material from the expedition, the Sherpa waved him away. This was, he said, the only time in his life when he had the opportunity of making a lot of money and he did not intend to ignore it.

When United Press quoted Tenzing on problems between the sahibs and their Sherpas, the fabric of the Hunt team was in tatters. There were repairs, instant, and others longer term, but the scars remained, particularly when various versions of the expedition and incidents on it were published. The under-current was plain, for instance, in the first Hillary reference to Tenzing in *Nothing Venture, Nothing Win*:

> I was eager to meet Tenzing Norkay . . . and I certainly wasn't disappointed. Tenzing really looked the part — larger than most Sherpas, he was very strong and active, his flashing smile was irresistible and he was incredibly patient and obliging with all our questions and requests. His success in the past had given him great physical confidence — I think that even then he *expected* to be a member of the final assault party on Everest as he had been with both the Swiss expeditions although I am sure that neither John Hunt nor any of the rest of us took this for granted . . . one message came through in very positive fashion — Tenzing had substantially greater personal ambition than any Sherpa I had met excepting perhaps Passang Dawa Lama who was also a mighty formidable individual. Tenzing would be keen to see us successful, that I felt sure, not that he had any particular

Hillary and Tenzing, garlanded with flowers and wearing the special commemorative medals presented by the Indian President, Dr Rajendra Prasad, to mark their climb.
ASSOCIATED PRESS

liking for the British but because this would mean that he too would be successful.

Even George Lowe, everybody's good guy, had qualifying comment on Tenzing. When it seemed briefly as if Evans and Bourdillon had gone on to the summit, Lowe wrote: 'Tenzing, we were hurt to find, lost his smile and did not share our enthusiasm. The idea of team effort had not been revealed to him and the idea that anybody but Tenzing should reach the summit was not pleasurable to him.'

But behind the Lowe criticism there was also admiration and affection:

> Tenzing was very fit; he moved beautifully and was well used to high altitudes. He was unspoiled, although even then (before the climb) he had received considerable publicity because of his climbing. He was a good companion who fitted in with everyone. He had an infectious sense of humour and a desire to yodel and whoop when he was happy . . . there is in him a natural elegance and a gentle manner. He is not a forthright character, but a dreamer and after Everest he was bewildered and often unhappy at the complications of fame. He has said, sadly, and I think honestly: 'I wish sometimes that I had never climbed Everest.'

George Lowe was right. Tenzing was completely out of his depth in the media-hype world he now inhabited. Clearly too he was not as happy either with the British expedition or Hillary as he had been with the Swiss and his assault companion Raymond Lambert. Raymond, a giant of a man who had lost all his toes after being trapped in a Swiss Alpine storm, had established a rapport with the Sherpa anyone else would have found hard to match.

Tenzing was honest in his preferences. He preferred the company of the Swiss and the French. He found the British more reserved and formal while the other Europeans treated him 'as a comrade, an equal, in a way that is not possible with the British. They are kind men, they are brave and they are fair and just always. But, always too there is a line between them and the outsider, between sahib and employee.'

He quoted how in Kathmandu on the way in, the Sherpas were allocated sleeping quarters in a garage, formerly a stable in the grounds of the British Embassy — the climbers were guests in the Embassy. There were no toilets and next morning some of the Sherpas used the road in front as a latrine. No one was happy about that. Inevitably, that story made the media; Tenzing denies he was responsible.

If Hillary was no Lambert — and you get the impression no one could have replaced the big, toeless Swiss — then the New Zealander was the next best thing. Tenzing's assessment:

> Hillary was a wonderful climber — especially on snow and ice, with which he had much practice in New Zealand — and had great strength and endurance. Like many men of action, and especially the British, he did not talk much, but he was nevertheless a fine, cheerful companion, and he was very popular with the Sherpas because in things like food and equipment he always shared whatever he had.
>
> I suppose we made a funny-looking pair . . . Hillary about six feet three inches tall and myself some seven inches shorter. But we were not worrying about that. What was important was that, as we climbed together and became used to each other, we were becoming a strong and confident team.
>
> One example of how we could work together happened while we were still on the icefall. Late one afternoon we were coming down, just the two of us, from Camp Two to Camp One, roped together, with Hillary ahead and myself second.
>
> We were winding our way between the tall seracs, or ice-towers, when suddenly the steep snow under his feet gave way and he fell into a crevasse. 'Tenzing! Tenzing!' he shouted. But fortunately there was not too much rope between us and I was prepared. Jamming my axe into the snow and throwing myself down beside it, I was able to stop his fall after about fifteen feet and then, with slow pulling and hauling, managed to pull him up again.
>
> By the time he was out of the crevasse my gloves were torn from the strain; but my hands were all right and, except for a few bruises, Hillary was unhurt. 'Shabash, Tenzing! Well done!' he said gratefully. And when we got down to camp he told the others that 'without Tenzing I would have been finished today'. It was a fine compliment and I was pleased I had done well.

When curiosity took me to a description of the same incident in Hillary's *High Adventure*, the comparison was significant:

> I didn't have much time to think. I only knew that I had to stop being crushed against the ice by the twisting block and I threw my cramponed feet hard against one wall and my shoulders against the other.
>
> Next moment the rope came tight and the block dropped away underneath me. Tenzing's reaction had been very quick. I cut my way to the surface without too much difficulty and thanked Tenzing for his capable handling of the situation . . . I found Tenzing an admirable companion — capable, willing and extremely pleasant. His rope work was first-class as my near-catastrophe had shown.

Sir John Hunt's official history of the expedition he led makes it clear that the un-Hillarylike description 'near-catastrophe' was justified:

Credit where credit is due — the head lama of Thyangboche receives the American Cullum Medal from Dr Charles Evans during a ceremony in 1954 when Sherpas from the previous year's expedition were presented with their Everest medals.
NZ HIMALAYAN EXPEDITION

> The whole mass of ice on which he landed collapsed beneath him and he fell towards a crevasse below. That no harm came of it was due to the foresight and skill of Tenzing, who was strongly placed against a slip on the part of his companion and held him brilliantly on the rope.

Compare the two differing memories. Tenzing: '. . . with slow pulling and hauling [I] managed to pull him up again . . . my gloves were torn from the strain . . .' Hillary: '. . . I cut my way to the surface without too much difficulty . . .' Their contrasting accounts did not end there but persisted right up to the Everest assault.

Hillary:

> I waved to Tenzing to join me. As he came down to me I realised there was something wrong with him . . . he seemed to move along the steps with unnecessary slowness . . . it was quite obvious that he was not only moving extremely slowly but he was breathing with difficulty and was in considerable distress . . . from the outlet of his face-mask were hanging some long icicles . . . the outlet tube was almost completely blocked . . . this was preventing Tenzing from exhaling freely and must have been extremely unpleasant for him . . . I was able to release all the ice and let it fall out . . . Tenzing was given immediate relief . . . my set . . . had partly frozen but not sufficiently to have affected me a great deal. . . .
>
> In front of me was the rock wall, vertical but with a few promising holds. Behind me was the ice-wall of the cornice, glittering and hard but cracked here and there. I took a hold on the rock in front of me and then jammed one of my crampons hard into the ice behind . . . I slowly levered myself upwards . . . I fought to regain my breath . . . my nerves were taut with suspense but slowly I forced my way up, wriggling and jamming and using every little hold . . . next moment I was reaching over the top of the rock and pulling myself to safety . . . I lay on the little rock ledge panting furiously. Gradually, it dawned on me that I was up the step . . . for the first time on the whole expedition I really knew I was going to get to the top. . . .
>
> When I was breathing more evenly I stood up and, leaning over the edge, waved to Tenzing to come up. He moved into the crack and I gathered in the rope and took some of his weight. Then he, in turn, commenced to struggle and jam and force his way up until I was able to pull him to safety — gasping for breath . . .

(In a *National Geographic* account, written in July 1954, Hillary said his struggle in the gap lasted half an hour. Interestingly, the magazine, in what purports to be a first-person account, quotes Hillary as telling Lowe: 'We knocked the *blighter* off.')

Tenzing's reaction in his book, *Sherpa Tenzing, Man of Everest*, republished in revised form in 1975, was:

> I must say in all honesty that I do not think Hillary is quite fair in the story he told indicating that I had more trouble than he with breathing and that without his help I might have suffocated. In my opinion our difficulties were about the same . . . we each helped and were helped by the other in equal measure . . .

On the Hillary account of the rock wall, now known as 'the Hillary Step', he writes:

> Again, I must be honest and say that I do not feel his account, as told in *The Ascent of Everest* is wholly accurate. For one thing, he has written that this gap up the rock-wall was about forty feet high, but in my judgement it was little more than fifteen. Also he gives the impression that it was only he who really climbed it on his own and that he then practically pulled me, so that I 'finally collapsed exhausted at the top, like a giant fish when it has just been hauled from the sea after a terrible struggle'.
>
> Since then I have heard plenty about that 'fish' and I admit I don't like it. For it is the plain truth that no one pulled or hauled me up the gap. I climbed it myself, as Hillary had done; and if he was protecting me with the rope while I was doing it, this was no more than I had done for him.
>
> In speaking of this I must make one thing very clear. Hillary is my friend. He is a fine climber and a fine man, and I am proud to have gone with him to the top of Everest. But I do feel that in his story of our final climb he is not quite fair to me; that all the way through he indicates that when things went well it was his doing, and when things went badly it was mine. For this is simply not true. Nowhere do I make the suggestion that I could have climbed Everest by myself; and I do not think Hillary should suggest that he could have, or that I could not have done it without his help.
>
> All the way up and down we helped, and were helped by, each other — and that was the way it should be. But we were not leader and led. We were partners.

Partners they undoubtedly were in the snow and ice of the peak, through the effort, the tiredness and the triumph. But in what followed and in what they wrote, that partnership ebbed somewhat and, I suspect, never returned. Understandably, others in the expedition, particularly John Hunt and Hillary, were shaken by the Nepalese public reaction and clearly there was a belief that Tenzing had, if not stirred it, then certainly enjoyed it. It is probably true too that the high altitude group were not

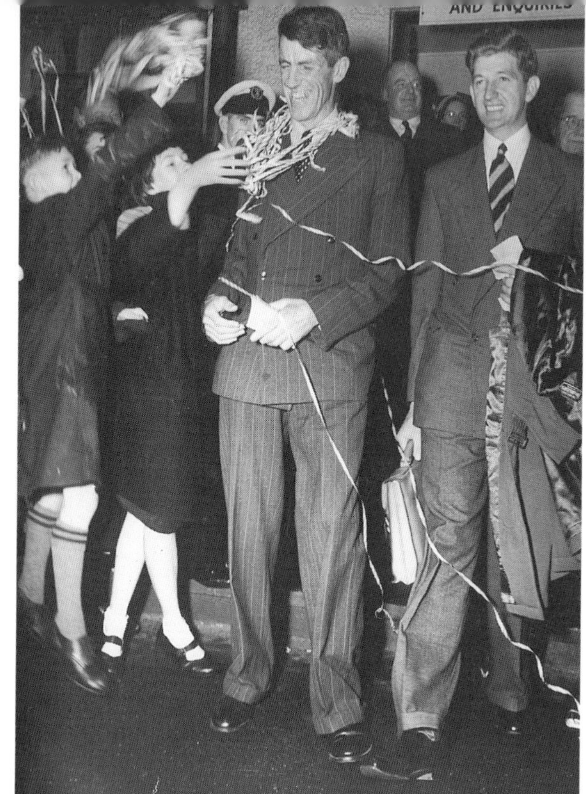
AND ENQUIRIES

in good shape to withstand this sudden change in political and personal altitude. Nearly a month's gap between the two knighthoods and Tenzing's George Medal did not help.

Of course, issues were put behind them as the months and then the years passed. Tenzing's own rather naive recanting, his clear statement that it had, indeed, been Hillary who had fractionally first gained the summit, his criticism of Nepalese nationalistic excesses and his almost abject re-acceptance of the Hunt team ethic — 'I would climb with him again' — may have helped bridge gaps between them which in those first days seemed as wide as the worst crevasses and seemingly as impassable.

He was loud in his criticism of that wave of post-Everest antipathy from the Nepalese.

> . . . I think that much harm has been done by narrow prejudice and nationalism, that Everest itself has been harmed and that my own people are at least partly to blame. The world is too small, Everest is too great, for anything but tolerance and understanding . . . whatever the differences between East and West, they are as nothing compared to our common humanity. Whatever the difficulties that arose about Everest, they are as nothing beside the common cause and the common victory, and to my English companions — to Hunt and Hillary and the others and all their countrymen — I reach out my hand across half the world.

Re-reading this in the context of forty years later, of lives lived and old events re-assessed, I was reminded of a discussion on Sherpa virtues by Peter Matthiessen in *The Snow Leopard*: 'When the going gets rough, they take care of you first. Their dignity is unassailable, for the service is rendered for its own sake — it is the task, not the employer, that is served. As Buddhists, they know that the doing matters more than the attainment or reward, that to serve in this selfless way is to be free.'

Temporarily over-run by the values and the pressures of a world he had not chosen to be part of, Tenzing had slipped off the high path, only to regain his balance and to once again find that freedom.

Top left: An arch of ice-axes greets Hillary as he comes ashore from the Tasman Airways Solent flying boat at Mechanics Bay in Auckland on his return home after Everest.
NZ HERALD

Top right: Once in the terminal, it was cheers, laughter and streamers for Hillary and George Lowe.
NEWS MEDIA

Left: Flags, speeches and a hero's welcome to Christchurch.
CHRISTCHURCH STAR

Perhaps it is too easy to pass a judgement, but it seems with that hindsight only the years provide that the problem was not Hunt, Hillary, the British or the Swiss. The starting point was the obviously deep and loving respect he had felt for and received from Raymond Lambert on those epic and unsuccessful climbs the year before. Even in the moment of triumph, he could not stop himself regretting that only Lambert's red scarf and not the wearer had gone with him to the peak.

In the same way, Hillary regretted that his assault teammate was not George Lowe. Lowe, who had first planted the seed so many climbs before, Lowe the step-cutter in the preparations for the ascent, Lowe who shared the great news first, who would climb with him and share the tough times on Makalu and then meet him in an historic moment at the South Pole in the years ahead. If anyone was, then he was Hillary's Lambert — the ever-present good friend to a man Lowe described with feeling to me forty years later as 'an ordinary but an extra-ordinary man'.

George Lowe, too, is far from ordinary. Like his mate Ed, he felt a childhood challenge to prove himself. An arm injured in boyhood skylarking had a doctor predicting that young George would be a cripple for life. Over George's dead body!

As a young man, he shared with his mate Ed a bubbling, good-humoured but, at the same time, deeply felt friendship. It was never more obvious than in those weeks of relaxation on their return from Everest. I remember watching one classic performance in the sitting room at the Hillary family home the day after their arrival back. Surrounded by family and friends — I had been invited in as a privileged outsider — he solemnly re-invested Ed with all the medals and decorations that had come his way. Ed stood on a chair for the occasion, seemingly in danger of knocking his head on the high ceiling. George Lowe pinned the emblems across his chest, down one leg and halfway up the other.

Then there was their schoolboyishly giggling account of the Buckingham Palace garden party. Resplendent in their Moss Bros hired gear, they strode down a path, twirling their umbrellas in masterful fashion as they went, until George Lowe's umbrella broke, coming apart at the handle and threatening to impale fellow guests as it flew off. 'Raw colonials swanking,' was his description of the incident again to me forty years later. Little wonder Hillary wrote of him: 'I have never laughed longer and louder than I did in his company.'

Lowe, the teacher, was a man of distinction. When he visited Repton, one of England's famous public schools, as a lecturer, he stayed on for four years, appointed there and then to the staff. He was for ten years headmaster of Grange English school in Santiago. His first wife was Susan Hunt, daughter of Lord Hunt, and the two men climbed together after Everest. A joint climb with Soviet Masters of Sport in the Crimea

Dressed for the occasion in hired suits, Hillary and Lowe stride out for the Palace and a royal garden party. Lowe twirled his umbrella in required fashion, only to have the handle come off in his hand and the rest of the brolly fly off like an unguided missile.
REUTER

in 1962 ended with the death of their Everest comrade Wilfred Noyce.

With Hunt, who was active in youth award schemes, including the Duke of Edinburgh scheme, George Lowe worked with youth. He went on a series of expeditions: to Greenland, on a walking tour of Greece, to Ethiopia, where he climbed the country's highest peak, and inevitably back to the Himalayas, where a group of young people planted 29,028 trees, a tree for every foot of one version of Everest's height. In South America, Lowe climbed the Andes. Of his three sons, Gavin, Bruce and Matthew, Bruce too has been to the Himalayas following in the footsteps of both father and grandfather.

Now, at sixty-nine, retired from years as an inspector of schools in Britain — interested predictably in outdoor education and geography among other subjects — Lowe lives ten miles from Derby with his second wife Mary; 'subsistence gardening' as he wryly describes it. And, to complete that long-ago link with a young Hillary in the back seat of a bus when he discussed his father's bees, tending bees and producing honey.

His links with Hillary are very real. After their Himalayan experiences, they shared lecture circuits together, he was the Hillary best man, son Peter's godfather, and they were at each other's sides to share the good times and the bad.

Lowe is a man of warmth and generosity. When talk gets back to Tenzing and the troubles, it is George who sums up: 'Poor old Tenzing believed all men were good, he was just so open and innocent.' He remembers the time at an investiture in New Delhi when Indian Prime Minister Nehru physically, spiritually and very meaningfully put his protective arms around the bewildered Sherpa. 'You are going to London and it is cold there,' the great Indian statesman said. He took off his long coat and put it around Tenzing's shoulders, like a protective cloak. (He did more than that. He later took the Sherpa to his personal suite and all but emptied a wardrobe to equip him for his travels. Everything but his famous trademark white hats.)

George Lowe saw this as more than a simple act of generosity. It was also a symbol of protection for someone who needed it very badly.

Forty years on, I asked the question: If he had his time over again, is there one peak he would climb? He did not hesitate. 'It could be Everest,' he said. Those were his words but it was easy to hear 'could' as 'would'. Then there was a qualifying silence.

'But, I'm glad it wasn't me. I was very close to it. I enjoyed all the pleasures and the excitement, but I didn't have to suffer any of the penalties as Ed has done, giving up totally his private life and being a public person ever since. He has done it marvellously but it is tremendous pressure and it is something I wouldn't have enjoyed or wanted.

'From that moment, he has been Mr Everest.'

CHAPTER FOUR

Dan Bryant's Legacy

Another piece of paper ripped off the cable network teleprinter. April 1954. Familiar names again. Makalu. Edmund Hillary. It is written into my memory — the greatest story I never wrote. Probably the deepest regret of my writing life is that I was not there. I should have been.

Hillary and I had become quite close after his return from Everest. My coincidental assignment as the bearer of the great news to his parents had earned me the role of the *Auckland Star*'s Hillary correspondent. Wherever he was, I was too. When he spoke, I reported him; when he lectured, I listened; when he expressed opinions or recalled events, I told whoever in the world was interested. And that seemed to be a lot of people. The Everest climber was New Zealand's number one celebrity, admired and sought after, listened to. Not just in his home country either. All the cable services seemingly could not get enough news about him. I found myself a key figure in this world-wide Hillary cottage industry.

Inevitably, with people like the young Ed Hillary and George Lowe at the centre of it — bouncy and good-humoured as they were, still buoyed up by the joy of that great event — familiarity added a deeper dimension to what could have been a rather boring and for them a sometimes irritating process. Instead, they treated me as a trusted near-friend. I could not have asked for more.

Just as inevitably, when Hillary told me of his plans to lead another expedition back to the Himalayas the following year, he agreed readily that I could travel with the party for the New Zealand Press Association and the *Auckland Star*.

It seemed as if I had been preparing for it all my life — I was then twenty-four. I convinced myself that a series of events in my life had led me to this point. There seemed reason to believe it. The year before I had fallen under the magical spell of the Himalayas through someone who knew and loved them well. He had passed the magic on to me.

The first time I met Dan Bryant, he was surrounded by the Himalayas. The great peaks were all around him, in photographs on his study wall. Behind him was a magnificent view of the north-west ridge of Everest and the description of what it showed flowed from him. Names I had never heard before — the Lhotse Ridge and the Western Cwm, and others, K2 and Kanchenjunga, Annapurna and Nuptse.

Dan Bryant had taken the photographs himself while with the Eric Shipton expedition of 1935. There, in his study, snow-walled with camera views of the great mountains but with New Zealand sun streaming in from green playing fields, he talked to me by the hour. It was a personal love story.

He had been a brilliant young scholar, an MA and a top honours graduate in history at twenty-one. Already a world-class mountaineer from experience on New Zealand climbs, he was invited to join Shipton on that

Headmaster Dan Bryant in his Pukekohe High School study where the mountains were all around him.
NICK BRYANT

A bearded Dan Bryant on Everest with the 1935 Shipton party.
NICK BRYANT

1935 reconnaissance. 'I have never been the same again,' he told me, and I believed him. There was a fervour when he talked about those mountains like a man irrevocably in love. He was.

Everest had escaped them that year. A series of thundering avalanches had closed off the northern approach they would have attempted. Instead, the party climbed anything else in sight — Dan Bryant reached 23,460 feet on one of ten Himalayan summits he climbed.

Shipton was not surprised. He had expected it of a climber who had come to him with great recommendations. In Europe for post-graduate studies, Bryant had climbed widely in the Swiss Alps, including a double traverse of the Matterhorn — from the Swiss slopes to the peak and down the Italian side, an hour's rest, up to the 14,700-foot summit and down again. All in fifteen hours.

Dan Bryant showed the same energy and skill on higher and technically more demanding peaks with Shipton. He had been home only a fortnight when the Shipton invitation came and he had willingly repacked his climbing gear and caught the first boat back to India.

It was almost as if Everest had reached out and touched me that day in 1952 in his headmaster's study at Pukekohe High School, just as it had done to him so much more deeply and permanently during those months in 1935. Neither of us knew then just how past and future were all around him at that time.

Dan Bryant talked reverently of the Shipton party discovering the frozen body of Maurice Wilson high on the North Col, of reading Wilson's diary in the battered tent. His last diary entry was that he would camp there 'just for a few days'. He never left the spot, and Bryant was one of the party who buried him where his attempt at a solo ascent had ended.

I can still remember the deep respect in Dan Bryant's voice as he recounted Wilson's journey into Tibet in disguise and what was known of his courageous bid to make history. What Dan Bryant did not realise on the high slope of the North Col, or as we talked about it seventeen years later, was that Everest history was with him as they left Wilson in a snow grave.

There with Bryant and the rest of the Shipton group was a young Sherpa called Tenzing in his first expedition to the mountain that would dominate his life. (Interestingly, the young Tenzing was, in his own words, 'angry and ashamed' at the failure of three Sherpas who had helped Wilson set up base camps, that they did not later search for him and had waited only three days for his return. They had breached Sherpa trust.)

The young Dan Bryant exchanged addresses with expedition comrades, including Shipton, and returned home. Back in New Zealand, he faced two attractive and differing invitations. One was to join Bill Tilman's successful attack on the 26,000-foot Himalayan peak Nanda Devi. It seemed an irresistible offer to the young Bryant. With Tilman in the conquest of the mountain they called 'The Blessed Goddess' was N. E. Odell, who had scoured the high slopes of Everest alone for Mallory and Irvine in 1924. Until the French scaled the taller Annapurna in 1950, Nanda Devi was the highest mountain yet climbed.

But there was another offer: for him to join the staff of Waitaki Boys' High School, one of New Zealand's most outstanding schools. He chose Waitaki. It was the beginning of a brilliant teaching career, at Waitaki, Timaru Boys' and Southland Boys', which took him to Pukekohe, then a small school but already, when we met, bearing the marks of Bryant's philosophy.

In an age where such views were revolutionary, he had abolished school rules, replacing them with 'just the dictates of common sense . . . every pupil and staff member accepting responsibility for the well-being of others and the school as a whole'. 'Functional authority', Dan Bryant called it. Pukekohe under Bryant — decades ahead of education theorists — had no class placings, just the pupil's mark and the class average. 'It does no one any good to know that someone is last in class.' A headmaster who was a sports enthusiast led a school with no sports championships or trophies, simply certificates for individual events. There were prefects but

Dan Bryant on a newly conquered 21,000-foot peak looks across to Everest, eighteen years before Hillary reached its peak.
NICK BRYANT

no head prefect. That would be announced much later in the year: 'an award for merit not a police appointment'.

These were the features of Dan Bryant the man which had attracted me to Pukekohe in 1952 to profile this distinctive and progressive headmaster (later killed in a car crash in 1957). I met the man I intended, but I also spent those hours, through him, in the presence of Everest, and another peak he pointed out to me with affection and respect. I had never heard the name before. It was Makalu. At the same time, Dan Bryant mentioned another name I had never heard before either.

My question was whether he had kept in touch with the men of the 1935 expedition. He certainly did, he said. He took a letter from Eric Shipton from his drawer. 'He's back in the Himalayas now . . . next year he hopes to attempt Everest . . . he wrote and asked me to recommend any New Zealand climbers who might interest him . . . there was one from around South Auckland who's already been in the area . . . he's got the right qualities so I sent Eric the name and he's with Shipton now. His name is Ed Hillary.'

I can still remember driving back to Auckland with a full notebook and a mind crammed with strange and exotic names, views of massive peaks, and Dan Bryant's imagery. He had talked about Hima laya, literally 'The Abode of Snows', Everest as Chomolungma, 'Goddess Mother of the World', as Tenzing called it later. He made magic of words I had never heard before but would again. Then, there were the other names.

Little over a year later, Everest had been climbed and Ed Hillary was an unforgettable name who had agreed on a proposition I could not turn down.

In the end, I did not have the choice. Shaken by the likely cost — a few hundred pounds as I remember — the Press Association decided that what would be a routine party to an area of the world which the Everest success had made so familiar did not justify the cost. They would not send me to Makalu. It was to me as if Stanley had been refused his steamer fare to Africa.

The decision was very minor in the general planning for the expedition done largely through a New Zealand-based committee. Both Ed Hillary and George Lowe, at first caught up with the aftermath of Everest, were home for only a few weeks before another avalanche of lectures overtook them. This time, they were a party of three. Ed had married Louise Rose, music student daughter (inevitably) of a former president of the New Zealand Alpine Club, and she accompanied them on their strenuous itinerary in Britain and the United States.

Hillary and Lowe were drawn back by a shared memory. It seems to an outsider as if the Himalayas have established their own web of people and events that entangles one and then another group in an intricate pattern of seeming coincidences which are far more than that.

It was there, so strongly evident, when I began my research, as apparently small pieces of personal history notched into others to form some unexpected jigsaw puzzle with a mass of high peaks as the central motif: Mallory and Irvine, who climbed 'because it's there', draw N. E. Odell into a solitary search for them on Everest; Maurice Wilson later dies in a lone attempt; his body is found by a party led by Eric Shipton and including Tenzing and Dan Bryant; Bryant recommends Hillary to Shipton, a nomination which is part of the process which sees Hillary on the summit — with Tenzing; Bryant is offered a place in the Tilman party — which includes Odell; Tilman later attempts Annapurna in 1950; with Tilman on that expedition is Bill Packard; when George Lowe and Ed Hillary organise their 1954 return to the Himalayas, Packard is a key figure in the arrangements in Britain; when Hillary later goes to the Antarctic as part of the polar crossing expedition, Lowe goes too with party leader Vivian Fuchs; with Hillary is New Zealand naval officer Peter Mulgrew; when Hillary and Lowe lecture in Auckland after their first Himalayan expedition, among the audience is a young schoolboy, Michael Gill; in 1960, when Hillary sets up a high-altitude research and climbing party on Makalu, Gill is a member; when Peter Mulgrew all but dies high on Makalu, Gill is one of the party that saves him . . .

It did not end there. Other pieces of the picture will fall into their obvious place as we go along. The choice of the party's area of interest was typical of that progression. After Shipton's bid on Cho Oyu was beaten back in 1952, Shipton, Hillary, Lowe and Dr Charles Evans from the party explored an area below the mass of Makalu, the Barun Valley.

Later, Hillary was untypically lyrical about the valley. This is very much how he described it to me: '. . . paradise . . . acres of red azaleas . . . waterfalls on sheer rock walls . . . rhododendrons in full bloom . . . grassy slopes clear of snow, carpets of primulas of every colour'. Not surprisingly, they believed it was a perfect area for climbing and exploration. And also no surprise, Charles Evans was one of two Britons who joined the New Zealanders.

There was no problem attracting candidates for this venture. The Everest success alone was enough to prompt what seemed like half New Zealand's experienced climbers and a good percentage of its novices to try for places. From them were selected Bill Beaven, Geoff Harrow, Norman Hardie, Jim McFarlane, Collin Todd and Brian Wilkins to join Hillary, Lowe, Evans and Dr Michael Ball, the English expedition doctor.

Focal point in initial planning was one peak Lowe and Hillary had been fascinated with from first sighting. In Hillary's words, 'it was a tall, graceful spire of fluted snow and ice'. They called it Baruntse. The plan was to climb one or several of 23,570-foot Baruntse, Chamlang, (24,012 feet) and Ama Dablam, (22,310 feet). The valley did not disappoint them at second sighting, but in the end they would remember it for more than its scenery.

The physical and mental memories of the recent Hillary lecture tour had gone with them to the high slopes. They had travelled hard, lectured hard and been entertained without respite for months. George Lowe describes them as 'fat and flabby . . . suffering from what Dylan Thomas once called "the ulcerous rigours of a lecturer's spring"'. Unconsciously, the two of them monitored the other's progress on the walk in to the Barun Valley, checking to see how the other refugee from the talk circuit was standing up to it all. This was the stage of any expedition that Ed Hillary liked least, as anyone who has been with him testifies. Nearly a year on from Everest, in April 1954, that reaction was more marked than ever.

For the Himalayan novices, this was new magic — the way new boys Hillary and Lowe had experienced it three years before, the sort of bonding Dan Bryant had talked to me about from his days with Shipton.

There was also something of a rebonding for the two veterans. Ed Hillary's fears the day that first note arrived on Everest addressed to him as Sir Edmund had already been borne out. In our exchange of letters about the Baruntse-Makalu party and when our paths crossed at stopping places on the Hillary New Zealand circuit (maybe circus would be a better description), he bewailed his fame. Once, he contrasted our business

Edmund Hillary at a rock and timber shelter on the lower slopes in the Barun Valley.
NZ HIMALAYAN EXPEDITION

Edmund Hillary and Geoff Harrow check out the day's lunch at the 1954 expedition's Barun Valley base camp.
NZ HIMALAYAN EXPEDITION

relationship with the dealings forced on him with what he called the 'Press vultures' overseas.

The man who had been invested by the Queen, lionised in London and New York, honoured by the King of Nepal, greeted in Auckland like a conquering hero, presented with the Hubbard Medal of the National Geographic Society by President Eisenhower in the White House, feted by Nehru in the magnificent palace the White Raj had left behind, found the peace and the beauty of the Barun Valley exactly as he had described it to me, as 'paradise'. There was not a Press camera crew within a thousand miles either. He made it clear before he flew out from Auckland that — while I shouldn't take it personally — he couldn't wait to get away from the questions and the flashbulbs.

I knew how he felt because I had in my minor way recognised the pressure too. Occasionally, at a particularly boring occasion, during a notably long-winded introductory tribute, I would catch his eye and he would wink — a conspiratorial gesture that told me plenty. Not that I could plead entirely not guilty. As Pat on the spot, I had answered the queries and the requests of what, at times, seemed half the world's media. In the process, I remember confiding to Ed then and later, I achieved something of a record. Not once, but three times in succession, his always rather tetchy father had ordered me from the Hillary property. Not that I left on any of those occasions, but the first time, at least, I was caught by surprise.

I had returned, 'bright-eyed and bushy-tailed', to the Hillary front door the day after my first call. Naively, and very unwisely, I had brought with me a glossy print of the photograph of the morning before, of Gertrude hearing the news. A thank-you gift for the family archives she was so proud of. The print never left me. Percy Hillary blasted me on first sight and pointed his intimidating finger towards the end of the front path. The picture of her with her mouth open, he made clear, had been an insult to his wife. He demanded that I go that instant. A promise of a second photograph, showing Gertrude in repose on the phone, and with her mouth precisely closed, published that night, placated him and I remained.

A photograph was at the heart of our next meeting, again the day after Ed's return from Everest. I called, asking for an opportunity to portray the hero in his home setting. That was impossible, said Percy, 'Sir Edmund' — I can hear his voice now, accentuating the title — 'is in bed'. The journalist in me brightened at this. Could we then, I suggested, get a picture of him in bed? At this, Percy Hillary's voice grew cold and the finger waved again. It would, he said, be highly improper to take a picture of 'a knight of the realm in his night attire' — his very words. At that, a familiar voice shouted down the stairs. 'For God's sake, Dad, let him up.' I went, the photographer with me, and we left with a shot of Ed in bed,

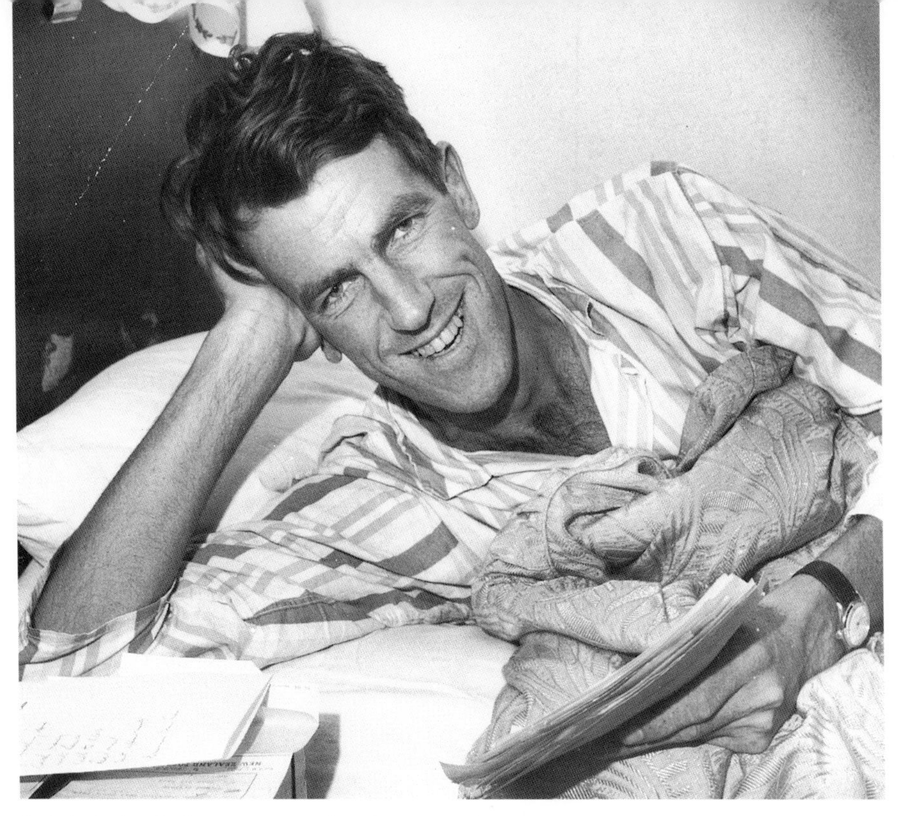

Knight in bed! The morning after his return home from Everest and his investiture, Edmund Hillary catches up with the first wave of letters that were to become part of his life.
NEWS MEDIA

resting on one lean elbow and reading welcome-home messages. I blush even now at my headline: 'Knight in bed'.

Percy didn't think much of it. He made that clear at our next meeting, which happened to be on the morning of Ed's wedding. Once again, Ed saved me and we chatted, more as I remember about the lecture tour than his wedding. It was 3 September 1953, Louise's twenty-third birthday. Next morning, they flew out.

That was, I think, the last day of the old-style Ed Hillary I had met first under a flurry of streamers as he and George Lowe arrived back by Solent flying boat at Mechanics Bay on a Saturday afternoon on their return from Everest. He was obviously overwhelmed by the size and the depth of the welcome, and that air of wonder stayed with him for weeks. It became indistinguishable from the joy he still carried with him from the climb.

When he came back for only a few weeks the next year, he was still the pleasant, approachable man who had left, but inevitably his celebrity life had added a new assurance; the pace of the tour had left him rather tired too, bemused by the last preparations for the Barun Valley trip.

None of this is a criticism of him. Times had changed. He had become a key figure in a world-size canvas and had been forced to adjust to that new role. He had done it well. Clearly, those first misgivings about the need for a new pair of overalls had widened far beyond that. Ed Hillary, Papakura beekeeper, had moved on. He had become a more polished version of what he once had been. The pressures on him were force-feeding in him abilities he may not have known he had before. He was now accepted as a leader of men.

He took this new persona with him to Baruntse, back to the environment he knew so well and which had been such a major factor in his metamorphosis. The country of the high peaks did not wait long to test him.

Sir Edmund Hillary relaxes in a special chair designed for him and presented by the people of Auckland on his return from Everest. Designer and maker George Emms of Gisborne told him the seat represented the Khumbu Glacier, an antimacassar was the peak and the lower right arm the adjacent peaks. It was specially intended for a particularly tall person who had a habit of slinging a leg over one arm of a chair.
NZ HERALD

CHAPTER FIVE

Crevasse!

For Jim McFarlane, then twenty-eight (now retired in Richmond, Nelson), that Barun Valley party was an overdue adult dream come true, but it turned very quickly into a long nightmare. He had been in the lineup when the 1951 New Zealand party was first talked about. But, like a lot of other climbers, money and circumstance ruled him out when the final quartet of Earle Riddiford, Ed Hillary, George Lowe and Ed Cotter headed off on the great trek that was the first stage of the Hillary Everest saga.

Jim McFarlane had the necessary record. He had teamed in a foursome of Riddiford, Norm Hardie and Bill Beaven on a number of outstanding climbs in New Zealand's Southern Alps. He had climbed Mt Cook and Tasman by new routes. He was a climber's climber — well thought of, respected, trusted.

With old companions Hardie and Beaven now in the Himalayas in 1954, he was determined to make the most of it, to compensate for lost opportunities three years before. Everything was a delight to Jim McFarlane. On one climb with Hillary and Brian Wilkins which had already given the three of them a magnificent vista of Everest, Lhotse and Makalu, he persuaded Wilkins to press on further for a first glimpse into Tibet. Hillary returned to await them at their camp.

McFarlane's glimpse of those peaks and plateaus was his last. He and Wilkins were returning on about forty feet of rope when disaster struck, or more correctly opened up. Wilkins was the first to break through thin ice over a crevasse and in a moment they had cascaded sixty feet into the darkness. Suddenly, there was just the silence and the knowledge that they had survived but were in a potentially dangerous situation. Wilkins was bleeding from cuts where his snowglasses had slashed his forehead. Jim McFarlane, obviously in pain, was dazed and could not move.

Sir Edmund Hillary looks towards the profile of Makalu from the base camp in the Arun Valley.
NZ HIMALAYAN EXPEDITION

Nearly forty years later, he has little memory of those first minutes although he was aware of Wilkins leaving to inch his way along the crevasse, taking hours of careful work cutting steps and slowly gaining height until at last he was at the surface. For Jim McFarlane, back in the crevasse, there was just the ache of his heavy bruising and the disorienting effects of what was later diagnosed as concussion. Consciousness and pain drifted around him. On either side, the high walls of the crevasse rose into the gathering darkness above, occasional cascades of ice and snow disturbed by their fall dropped around him.

It could have been minutes, in fact it was hours, before Brian Wilkins worked his way back, stumbling at times across the rough terrain to the base camp where Ed Hillary waited, at first rather angry at their late arrival and then increasingly concerned for their safety. First glimpse of the bloodied face of Wilkins, tense with strain and tiredness, confirmed his worst fears. There was just enough time for Wilkins to tell his story before Hillary headed back with him and a party of Sherpas.

The first problem was to find the hole above the crevasse. The light was going very quickly and the journey back to the accident scene was across unstable rock. Suddenly in the now poor light, they sighted Wilkins' hat where it lay on a boulder, and signs of his tracks.

For Jim McFarlane, the first indication that he was no longer entirely alone was Hillary's voice calling to him down the surface hole in the glacier. For what must have seemed hours to those listening sixty feet above, the injured man struggled for the strength to reply. In that delay, both Hillary and Wilkins feared a real disaster. When McFarlane's response came it was weak — but a reply. He was still alive and comparatively conscious.

Nearly forty years later, Jim McFarlane can still remember the light of Hillary's torch around the iced walls. If McFarlane's danger continued, so did that of his rescuers. The edge beneath Hillary's body as he swung the torch and inched forward trying to glimpse what lay below was thin and fragile. The risk was that it might break away, plummeting Hillary to the floor where McFarlane lay and perhaps burying both of them in ice and snow in the process.

The Sherpas too were very unhappy about the situation, and said so. As intent as anyone to recover McFarlane, they were also conscious of the possible threat to the safety of all of them. When Hillary told them he planned to descend into the crevasse on a rope in an effort to bring McFarlane up, their concern deepened. As McFarlane put it those years later, they 'took fright'.

Knowing that time was their enemy and increasingly worried, Hillary began his slow drop into the crevasse, on one rope and taking another to bring back McFarlane. The Sherpas were emphatic he should not take the risk. He saw no option. Their doubts and their very real concerns about the permanence of the lip they were on clearly affected the Sherpas. The rope around Hillary's chest began to tighten painfully against him but he was determined that he would withstand the pain until he reached McFarlane on the ice floor below.

Finally, it was clear that he was not going to make it. McFarlane could just make out the pendulum shadow of Hillary swinging on the rope above him, but not getting any closer. For some unexplained reason, the Sherpas had stopped lowering him and he was left hanging in the space between the narrowing ice walls, in real pain from the pressure of the rope and getting nowhere.

There was now only one way to go — up. When Hillary began shouting to the Sherpas to raise him, Jim McFarlane can remember adding his thin voice to the demand, shouting 'Uppa, uppa', watching the shadowy figure above him disappearing and wishing it was him. For his part, Hillary would cheerfully have changed places with anyone — except McFarlane. The rope was causing him real problems and pain while he became trapped under the overhanging edge of the crevasse. Now in a real state of panic, the Sherpas merely pulled harder. The rope tightened even more and the overhang seemed impassable.

By the time Hillary finally levered his way past the obstruction and over the edge to safety, he was in severe pain himself. He had, in fact, broken three ribs and the extent of the effort at such high altitude had left him struggling for breath and exhausted.

It seemed plain that more manpower might be needed, and light too, to get McFarlane to the surface. First plan was to lower sleeping bags to make him comfortable for what was certain to be a night in the crevasse.

With that rope down to him, it seemed logical to maybe make one more effort. Told to tie the rope around him, the injured man finally achieved that and called up for the party above to begin the lift. But only so far. Once again, the overhang that had proved such a problem for Hillary was more than a weakened and injured McFarlane could cope with. At one stage, his hand seemed within grabbing distance but neither he nor Hillary could do more than briefly touch fingers.

How did he feel as the rope began to lower him back where he had come from? Jim McFarlane can't remember clearly through, first the pain, and then the years. 'I probably thought "bugger it".' With no option now but to leave him where he lay, Hillary and Wilkins tried to assure themselves that he was as comfortable as was possible in the grim surroundings below.

McFarlane can remember vaguely their instruction that he get into the sleeping bags and his assurances that he had done so. In fact, he hadn't. In his stunned condition, he thought he had but had instead

A line of Sherpas, barefoot in the snow, making their way in towards the high camps.
NZ HIMALAYAN EXPEDITION

merely wrapped them loosely around him. It would not be enough to keep out the intense cold of that long night.

Later, he told me: 'That was one thing, but the stupidest part of it was that I went to sleep. Maybe it was the concussion too. But they say that's one thing you must not do. Sleep lowers your body temperature which stays hotter while you are awake. One thing, the concussion acted rather like an anaesthetic. Things didn't seem quite as serious to me as they really were.' It was probably just as well.

Regretfully but inevitably, Hillary and Wilkins returned with difficulty to their tent, leaving the rope to Wilkins tethered.

It took only a second after waking early next morning for Ed Hillary to realise two things: he had been injured in the abortive rescue attempt of the night before — his chest was very painful — and there was real need to act quickly to save Jim McFarlane. It was already snowing and they had to head off any serious change in the weather.

If, indeed, McFarlane had survived the night. They were confident that he had, but at that height and in those circumstances, you could not be sure. Had they known the real state of affairs down the crevasse, they would have been even more concerned.

When Brian Wilkins retraced the escape route he had taken from the ice trap the night before, the news he brought back was not good. Jim McFarlane had not been in the sleeping bags overnight. Worse than that, he had for some concussion-affected reason taken his gloves off and his hands were ominously frozen and stiff. His back was giving him trouble and he had great difficulty moving. Wilkins had left his companion in a sling harness ready for the lift to the surface, which was getting increasingly urgent.

But, once again, that ledge blocked the rescue. For the second time, they had to literally painfully, slowly lower McFarlane back to the ice floor below. Carefully, as a matter of life and death, Brian Wilkins and the Sherpa Da Thondup chopped away at the lip of the crevasse, tiny piece by tiny piece to stop a major ice fall onto the victim below.

One more attempt and McFarlane was almost there, inevitably jammed but within reach of the Hillary hand, which grabbed part of the sling and pulled the injured man free of the obstruction. Did the always cheerful and resilient Jim McFarlane have a special one-liner for his rescuers? Thinking back through the years, he doubts it. 'I only remember coming to the surface . . . I was just too poked to say anything memorable'.

He carried the marks of his ordeal thus far and the pointers to months of pain and problems ahead. His hands were frozen stiff. He says those fingers that remain are still rather clawlike. Inside his boots, his feet were icehard and without feeling or movement.

For his rescuers too, the problems were simply compounding. They made up an improvised stretcher out of pack frames and began a shocking descent towards their camp site, five Sherpas with either Hillary or Wilkins in turns struggling sometimes only a few feet at a time before the thin air and the sheer energy-sapping effort of it forced them to rest.

With it all, there was the other almost unrecognised factor at that stage, that Hillary was himself injured and in great pain. Realising that they faced a task which was quite beyond them to get McFarlane to the medical help he so urgently needed, the party decided instead to bring their own forward camp to him. Wilkins remained with the injured man as the others moved their tents and supplies to them and while Hillary and a Sherpa went for help.

Despite his injury, Hillary drove himself as only he could. But the intensity of the pain in his chest and the effort of the descent across badly broken surfaces forced the pair to rest overnight before reaching the base camp across a river early next morning. Once again, Hillary's own account of the arrival carries one of those significant asides: 'George Lowe's strong, confident figure came towards me and I felt a lifting of my burden.' George who was always there . . .

The combined efforts of Sherpas and sahibs took a further four days to bring Jim McFarlane down the glacier to the base camp. He was as cheerful as ever but clearly in a bad way and in need of prolonged and expert medical care.

The two doctors in the party, Michael Ball and Charles Evans, were both emphatic that he should rest and then go out of the mountains and back to New Zealand.

Jim McFarlane has little recollection of what was obviously a hair-raising descent to the base camp, but he does remember the three to four weeks he spent there recuperating and preparing for the trek out to India. Both he and Mike Ball became conscious that the damage was serious. Both hands and feet turned black and the flesh began to rot away. The

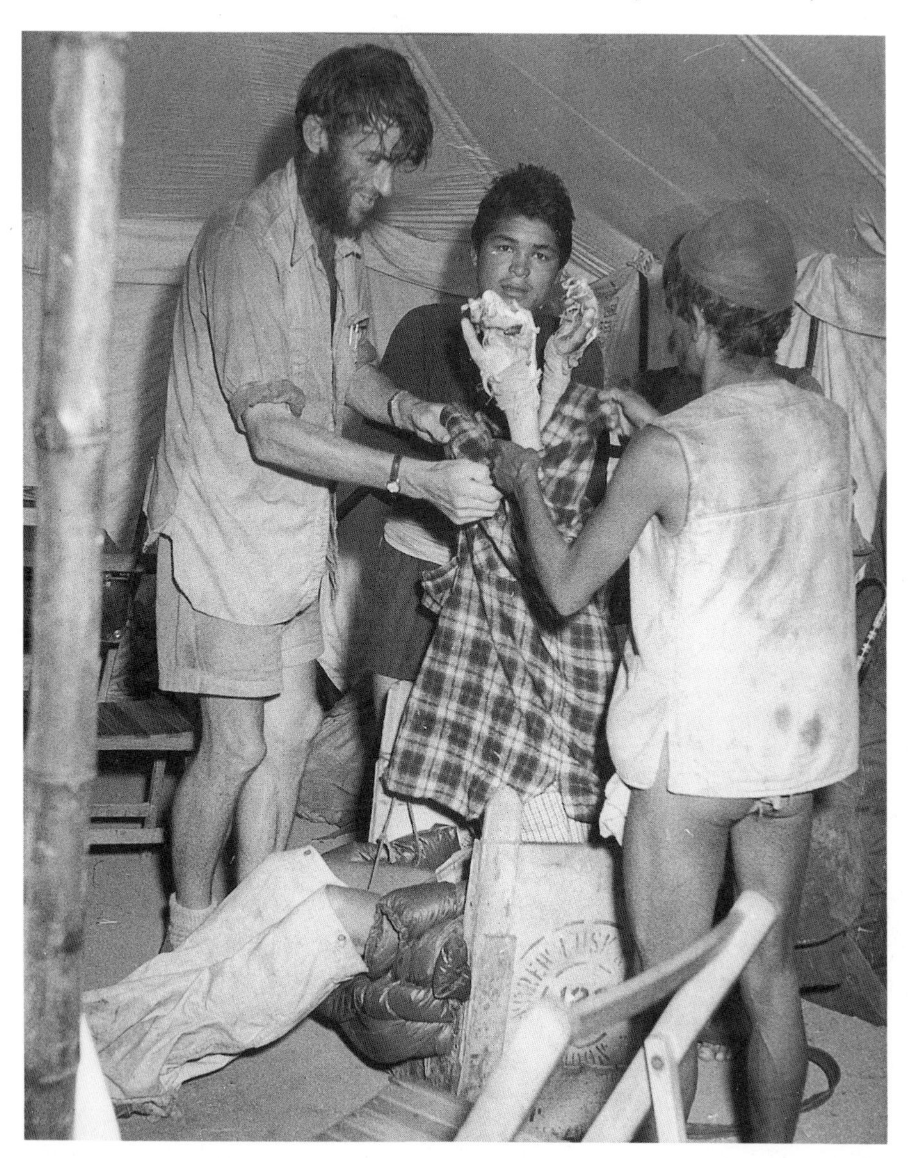

Edmund Hillary helps dress Jim McFarlane. In the centre of the photograph is McFarlane's badly frost-bitten hand.
ASSOCIATED PRESS

rest stage was made easier by daily visits from members of a nearby Californian climbing party who planned an attempt on Makalu. 'They would come over and we would send down the valley for a four-gallon can of Sherpa booze and spend the whole day reminiscing and disposing of the stuff. Charles Evans, the doctor, reckoned it was quite the best thing for getting circulation back into limbs affected by frostbite.'

Jim McFarlane laughs about it now and remembers the good times, but the worst times were obviously very bad.

He was not the only one in trouble. The one characteristic everyone attributes to Ed Hillary is that he hates being bored and, as his fellow climbers put it, 'sitting around on his bottom'. That certainly applied in the weeks after Jim McFarlane's crevasse fall. The pain was still there in his chest, but after waving away a reconnaissance party planning to push up Makalu, Hillary became restive. Charles Evans had reached 23,000 feet and things looked promising.

The combination of that news and Hillary's natural and oft-quoted reflex about doing nothing proved too much. He headed up the mountain 'much too soon for his own good' as George Lowe describes it. The Hillary words: 'I couldn't bear to be left out and although my ribs were still troubling me I decided to join the assault.'

It soon became very clear, even to Hillary, that he was in no condition to be where he was, attempting what he was. One morning, he fainted putting on his boots. An hour later, he was in what normally unflappable Lowe defines as 'a serious condition . . . delirious . . . unable to recognise anyone in the expedition but me'.

Hillary himself remembers shocking hallucinations of clinging to the ice cliffs of Makalu, being pounded by avalanches with other climbers screaming for help. He was in a very bad way. George Lowe remembers his friend's ramblings including — shades of McFarlane — constant fears of frost-bitten feet.

So they put together another make-shift stretcher to carry another difficult load — this time Hillary himself, a three-day task for the Sherpas. Again, George Lowe makes no bones about the danger. With oxygen available — there was none on the upper slopes — Hillary began to improve but it was touch and go. He had a form of pneumonia. One further complication was that he had the unlikely symptoms of malaria, which puzzled Charles Evans until Hillary was able to recall that he had had malaria in the Pacific. By now, the party had two members heading down the mountain.

Behind them, on the high 23,570-foot spire of Baruntse, the peak whose grace and challenge had first taken Hillary and Lowe back into the beauty of the Barun Valley, drama and excitement went on. Makalu was behind them, the McFarlane-Hillary incident had swept away any pros-

Edmund Hillary carried out of a tent at Camp Two at 19,000 feet on the Makalu Icefall after collapsing with pneumonia and malaria.
NZ HIMALAYAN EXPEDITION

Hillary on a stretcher (foreground) as his bearers rest on the long carry back to the base camp.
NZ HIMALAYAN EXPEDITION

Relaxing between big adventures. When the New Zealanders produced their inevitable rugby ball soon after reaching the Arun River area, Sherpas were quick to join in their own version of the game.
NZ HIMALAYAN EXPEDITION

pect of further advance there. Makalu would keep for another day — and, unknown to them, another high-level crisis.

Instead, before Hillary left on his disappointed trek out of the Himalayas, he and Lowe had agreed that the party would attempt previously unclimbed Baruntse. Lowe, Beaven, Todd and Harrow set off with Sherpas to position themselves for the assault. On the high slopes, at around 20,000 feet, the Sherpas staged what must have been the first rugby game ever played at that altitude. Inevitably, the New Zealanders had brought a football with them, and having had their enthusiasm fired by an earlier demonstration lower down the valley, the Sherpas staged their own rematch. Years later, Graeme Dingle was to star in another game. In 1954, in this Baruntse natural stadium, the Sherpas had the contest to themselves. The All Blacks would not have been deterred by the technique, but would have struggled in the setting and its thin air, the rugged ice ground.

Cold and wind followed. Baruntse may have an air of aloof beauty, but its challenge also called for very precise and consistent skills.

Later, Lowe and Beaven would tell of a near disaster. They were working their way along a ridge in strong wind when suddenly the surface began to move, sending part of it crashing down and leaving Beaven literally teetering on the very edge, 'waving his arms gently like a tight-rope walker to keep his balance' was how Lowe later described the scene.

Later that day, Todd and Harrow pressed on to the summit of Baruntse in snow and high wind, then returned in rapidly failing light down the ridge where their two companions had nearly come to grief, met by an anxious Lowe and the Sherpa Mingma in total darkness and in worsening weather. The descent had been a nightmare, but the two climbers' complete control of technique in the most difficult conditions had conquered Baruntse and returned them safely.

Mallory, that legend who has come to epitomise so much about the Himalayas and people who challenge its peaks, summed up decades before: 'Have we vanquished an enemy — none but ourselves.' There were no thoughts as profound or as lyrical as Mallory's when wind and snow lashed the tent high on Baruntse that night, simply pride and relief — plus, for Beaven and Lowe, an eagerness to get ahead with their attempt.

Two days later, Baruntse was theirs as well. But, again, not without danger. Working on a long rope and in weather that began well but deteriorated into a genuine storm, high winds and driving snow, they made their descent, hour after hour of effort and tension, knowing that the slightest slip would have been the end of them, in cold so intense that Lowe had three pairs of gloves freeze. The last of the descent, as for Todd and Harrow before them, was made in complete darkness.

Baruntse had been a challenge — and they had met it, now simply one of the nineteen peaks over 20,000 feet the party had climbed.

Meanwhile, Hillary and McFarlane were on their long journey back to the railhead. The party had made a carrying seat for McFarlane out of a wooden supply box, and Sherpas carried him on their backs, twenty minutes at a time, across seemingly frail swing bridges, down precipitous paths and difficult rock faces, day after day. Again typically, Jim McFarlane jokes about it.

'I had quite an enjoyable trip actually . . . my clearest memory of both the stay at the top and the trip back is the amount of Rakshi, the local grog, we drank. It's a distilled liquor not unlike kerosene but very potent stuff.

'The porters, in particular, at their midday stop would stoke up and it was sort of 'drunk in charge of a patient' after that but they all seemed pretty steady on their feet.'

No word of complaint about what must have been in so many ways a horror trip, no recall of the pain, the sense of despair and very real concern about what lay ahead for him. There was only total admiration for the Sherpas who carried him out, struggling over barely formed tracks and across rough slopes with their unwieldy load on their back, McFarlane constantly bumped and swaying, no doubt conscious of his injuries and aware of the non-stop risks as they went.

Twenty days later, he had his wounds dressed at an Indian hospital overnight, was given new dressings and lots of eau-de-Cologne and piggy-backed by stewards onto an airliner for the long flight home from Calcutta.

'My feet and hands were quite rotten at this stage and they dressed me up for the journey. Unfortunately, the flight was delayed and the eau-de-Cologne wasn't quite strong enough. By the time I got through Singapore and Indonesia I was really stinking to high heaven.

'What stuck in my mind was the forbearance of the other passengers at this mixture of rotting flesh and Cologne. No one wrinkled their nose.'

Back in New Zealand, McFarlane faced amputations of half of both feet and his little fingers from rather clawed hands with months of plastic surgery and post-operative treatment. But he went on to tramp and sail, build houses and continue his life.

He talks now without the slightest rancour about both the beauties of the Barun Valley and the deadly bite that lies behind that outward appearance. In particular, he is grateful for the opportunity of meeting and working with Hillary, described in the typical understatement of climbers as 'a very useful citizen . . . I feel privileged to have known him'.

Of those weeks long ago in the shadow of Makalu, he is relaxed in

Jim McFarlane is borne down the Barun Valley by a Sherpa carrying a chair made from a box after his crevasse accident at the head of the Barun Glacier.
NZ HIMALAYAN EXPEDITION

his memories and in his humour. 'You should have come,' he says. 'You could have made a lot of money for them, what with people falling down crevasses. There was a facetious comment made that such expeditions should have one or two expendable members to throw off cliffs and so on to get a bit of publicity. I think that inspiration came from me.' For Jim McFarlane, this was his last climb.

One of the bamboo and timber bridges built to explore valleys on the far side of the Arun River. Later, these were used by Sherpas carrying out the injured Jim McFarlane.
NZ HIMALAYAN EXPEDITION

For Ed Hillary, it was a first hint — the rugged, ever-strong, ever-resilient man of Everest had experienced for the first time a clear indication of the frailty that is within all mankind. There had been other setbacks in the past, but none as public as this. News of his condition and danger had raced around the world. That Reuters message on my desk was simply part of an avalanche of fact and theory about him.

The Times of London discovered at about the same time as that message reached me that they had no obituary prepared on him and tried to convince Sir John Hunt he should provide one. Hunt, shaken and at the same time unconvinced that someone so seemingly indestructible should be at such risk, suggested it wouldn't be necessary. He was right. But, like the tribute *The Times* no doubt prepared anyway and put into storage — into what newspapers morbidly call 'The Morgue' — something lingered on after that first incident on Makalu.

It was not necessarily in Hillary's mind, for he showed absolutely no inclination to duck danger or shirk risk afterwards, but perhaps somewhat less obviously in his physical makeup. As events would later show very vividly, he would find increasing problems in withstanding the effects of the high altitude in which he had once been the unchallenged master.

As if in compensation, the younger Hillary with those apparently inexhaustible supplies of energy, drive and strength, was giving way to a new character, absorbing the lessons of human management and planning from such people as Sir John Hunt. He was now able to direct as well as inspire, plan as well as lead by example, still unconsciously wanting to be in front as his competitive nature demanded, but also able to be that crucial one step ahead in strategy too.

The Himalayas tend to inspire people like Mallory to poetry. Ed Hillary is not one of them. That's no criticism. His prose has always remained straightforward and direct, like the man. But, by the end of 1954, more than a year after those fifteen minutes on Everest, and even months after his recovery from the effects of the McFarlane rescue and his ill-judged bid to be part of the Makalu assault, he was hearing something of what Himalayan veteran Bill Tilman had once said.

Maybe it was about the time Tilman invited a young New Zealand schoolteacher called Dan Bryant to join him on Nandi Devi. Whatever the date, what Bill Tilman said then, now had significance in the life of Ed Hillary: 'We live and learn — and big mountains are stern teachers.'

CHAPTER SIX

The Call of the Pole

Hindsight is a wonderful thing. With that advantage, Hillary's link with the South Pole seems so obvious, you would wonder that anyone would be surprised at the events of 1957–58 on the southern polar ice. It was somehow inevitable.

Robert Falcon Scott was to the South Pole what George Mallory was to Everest — the legendary figure swathed in both heroism and mystery, the epitome of courage and honour and crowned with the ultimate acolade, glorious failure. The young Ed Hillary was one of a generation brought up on the story of Scott in a country with long-standing links with South Polar exploration.

Captain James Cook, who claimed New Zealand for the British Empire, also voyaged south to the polar ice. 'Lands doomed by nature to everlasting frigidness and never once to feel the warmth of the sun's rays, whose horrible and savage aspects I have no words to describe,' he wrote. A century and a half later, New Zealand ports harboured now long-dead explorers, crowds waved them off to the ice, cheered their return, and, in the case of Scott, mourned their loss.

School libraries like mine in Hawera had featured an icon of our times, the coloured painting of a man called Captain Oates leaving a wind- and snow-blown tent, his arm protecting himself from his final blizzard. The polar ice had represented for so long one of the last challenges. Who could have doubted that with Everest conquered, not only by Hillary and Tenzing but by the Swiss too in 1956, the southern ice would call Hillary next?

Like Mallory and Everest, the southern venture was a link with heroes and history. The basic component was a British plan led by Vivian Fuchs to cross the Antarctic continent from the Weddell Sea.

This was the great trek Ernest Shackleton sought to achieve forty years earlier. It called for a journey through the South Pole, which had drawn Roald Amundsen in December 1911, and then on across the waste of ice where Scott and his party had died in their retreat from the Pole a month later.

Knowledge of the history is essential in allowing us to see the full significance of what was planned and what finally eventuated. In so many ways the history is the tragic story of Scott, his life and death, what he attempted and his final fate, what the manner of his dying came to mean for generations. It is also the achievements of men like Shackleton.

As with so much in formal history, a Maori legend predates provable fact; Ui-te-Rangiora once sailed south in a large waka (canoe) into the frozen ocean.

For the late Victorians and Edwardians, it was men like Scott and Shackleton who made the southern ice a place of mystery and courage. After a Belgian crew in the *Belgica* had proved it was possible to survive on the continent — they were trapped in the pack ice of the Bellingshausen Sea for a year between 1898 and 1899 — Scott and Shackleton led three parties in exploration on the ice. The first was Scott's *Discovery* expedition between 1901 and 1904, and the next was Shackleton with *Nimrod* for two years from 1907, when he and a party of five reached a point ninety-seven miles from the Pole.

From all this developed a rich literature on polar exploration, not the least the Cherry-Garrard account of an epic and terrible trek to Cape Crozier made by a British group from Scott's *Terra Nova* party in 1911 to visit an emperor penguin colony. The title told it all: *The Worst Journey in the World*. Place names on the route remain as a vivid memory of that journey — Terror Point and Mount Terror. (A young New Zealand climber called Hillary had always rated that book his best adventure reading, plus *Camp Six* by Frank Smythe, the Englishman who reached 28,000 feet in an assault on Everest without oxygen, and *Nandi Devi* by Eric Shipton. Significant reading.)

Both Scott and Shackleton were innovators, and that too was an aspect of polar history that carried through into the events of the 1950s. Scott used a captive balloon in pioneering style as aerial reconnaissance as early as 1902. Shackleton used a motor car at Cape Royds on Ross Island in 1908 and, later, Manchurian ponies in an exploration of a route up the Beardmore Glacier.

When Scott attempted the Pole in 1911, he sought to use a combination of traditional dog sleds, plus motor sledges and ponies, with tragic consequences. The engines broke down very quickly, the ponies had to be shot and the dog teams were sent back as Scott and his four companions slogged for the Pole dragging sledges. After a battle of more than eleven weeks, they reached the Pole on 17 January 1912, only to discover the flag of Norway already flying there. Amundsen had beaten them by a month, arriving on 14 December 1911.

Amundsen had been first mate with the Belgians who had earlier wintered over. Originally aiming at the North Pole — reached by Robert E. Peary in 1909 — he continued preparations, borrowing a ship from another explorer, Nansen, and sailed, most people thought, for the North Pole. He turned secretly south, to set up a base sixty miles closer to the South Pole than Scott and make the journey with four companions and fifty-two dogs.

Scott's sighting of his rival's symbol of success — a Norwegian flag at the Pole — must rank with the most crushing disappointments in history. After such effort, such disaster.

Worse was to come. On their heart-breaking return journey back down the Beardmore Glacier, they were constantly battered by appalling weather. One man, Edgar Evans, died on the glacier a month after leaving the Pole. A second, Captain 'Titus' Oates, chose death rather than further threaten the hopes of the remainder. Crippled by frostbite and unable to keep up with his companions, he walked out of their tent into a blizzard on 17 March 1912. His farewell became his epitaph, like Mallory and the mountain. 'I am just going outside and may be some time.' He was, as Captain Scott described him in his journal, 'a very gallant gentleman'. That was the title on that painting in my school library and in hundreds of others.

On 29 March, trapped by yet another blizzard and only eleven miles from a depot, Captain Scott wrote his last entry in his diary, wrapped himself in his sleeping bag and prepared for death. With him was H. R. 'Birdie' Bowers and Edward Wilson, the pair who had battled through what was until that moment 'the worst journey in the world'. Their bodies were found in their frozen tent on 12 November 1912. In my archives is the page of the *Auckland Star* that told of the discovery, including Sir Ernest Shackleton's awed reaction: 'He would have died hard.'

For more than forty years, an overland journey to the Pole — perhaps as part of a crossing of the continent — remained a dream, always coloured by the fate of Scott and the near-disaster when Shackleton attempted the feat in 1914 only to have his ship, the inaptly named *Endurance*, crushed in the pack ice of the Weddell Sea. What followed must rank with the greatest feats of human courage and endurance in history. Shackleton and his party and crew had at first drifted for 281 days and 1500 miles on their crippled ship before they were forced onto the ice floe. They carried lifeboats from the sunken ship on to a solid floe before drifting on it to bleak Elephant Island four months later.

Shackleton, with a crew of six, then sailed 800 miles in an open boat, through murderous gales and waves, to bring back a rescue ship to break through ice floes and save the remainder, who had survived by living under their upturned boats on the shore.

Shackleton died of a heart attack while sailing yet again to the Antarctic in 1921 and was buried on South Georgia, where he had arrived in that open boat six years earlier.

Amundsen, like many before and after him, not the least of them Hillary, lived out a full life of adventure, including flying over the North Pole in the dirigible *Norge* in 1926. He died flying to rescue the Italian engineer and polar flier Umberto Nobile, who had flown over the North Pole with him and two years later crashed another airship on the ice near Spitsbergen. Nobile survived but seventeen others were killed.

Inevitably, with the tremendous advances in aviation in the 1920s, the Antarctic lost something of its isolation. Hubert Wilkin's first flight on the ice continent in 1928 was followed by the American explorer Richard Byrd's flight over the South Pole in November 1929. Byrd had flown over the North Pole three years earlier. It was logical that Byrd, by then an admiral, should return in 1946–47 with a typically massive American air and sea fleet, 4,700 men, thirteen ships and twenty-three aircraft in Operation Highjump.

The Antarctic would never be the same again. By 1962, the United States had also dragged the area unwillingly into the nuclear age with the setting up of a disastrous nuclear power plant at its McMurdo Sound base. The plant had a history of fire, radiation leakages and other problems until it was decommissioned and withdrawn ten years later, involving the removal of 390,000 cubic feet of contaminated rock to mainland America for disposal.

Science and technology had by the mid-1950s already begun to play a dominant role on the ice. Facilities that research offered gave new strength to the Fuchs plan of a trans-polar crossing. There was irony in the beginnings of the Fuchs saga. Shackleton's ship had been crushed by the polar ice in 1914. The Fuchs journey began in his mind while he was also trapped by the same ice pattern forty-five years later. But that is to jump ahead rather too many chapters in the life and times of Vivian Fuchs.

He was born on the Isle of Wight in 1908, son of a German who settled in England and married an English bride. Vivian, who somewhere along the way picked up the nickname of 'Bunny', spent his early childhood in a Britain dominated by anti-German feeling, where even to own a dachshund was bordering on the unpatriotic. Friends and acquaintances later wondered whether this had contributed to a somewhat withdrawn personality, overlaid with what at times seemed almost an unconscious send-up of British stiff upper lip and understatement.

He studied geology brilliantly at Cambridge, leaving with a doctorate in time to join an expedition to Greenland, went with a Cambridge university party to the East African lakes in 1930–31 and was leader of an expedition to the borders of Ethiopia and Kenya, to the Lake Rudolph Rift Valley and then another to Tanganyika before war service as a major (he was mentioned in dispatches).

In November 1947, he was appointed leader of the Falkland Islands Dependencies Survey in Antarctica, mapping and collecting meteorological, geological and biological data. In 1949, the ice that smashed Shackleton's dream played a part in producing another.

Fuchs and his party were left on the polar coast when the relief ship *John Briscoe* could not penetrate thick ice 250 miles short of their base and had to abandon them for the winter. On one dog-team journey, a thousand-mile round trip from their base at Stonington Island, blizzard-bound in his tent for three days, Dr Fuchs conceived the idea of doing what Shackleton had planned, a journey across Antarctica from the Weddell Sea coast to the Ross Sea.

He lit and vigorously stoked the fire that was to become the Commonwealth Trans-Antarctic Expedition, and not surprisingly, in 1955, was seconded to lead it. One of the strongest influences in his acceptance of the role and in his planning was Sir James Wordie, Master of St John's, his old Cambridge college. Wordie was chief scientist with Shackleton, on the ill-fated *Endurance*, and the man who had, in turn, led the young Fuchs on his first student journey to Greenland so many years before.

Once again, George Lowe was the catalyst introducing Ed Hillary into the pattern of events. Fuchs had already approached Lowe to join his party on the strength of his impressive photography in the Himalayas. Lowe's role, it was explained, was to be official photographer and interpreter, to make sense of New Zealand messages, Fuchs explained with a laugh. In the years that followed, there would be times when observers wondered whether Fuchs was unconsciously reading the future!

Fuchs' impressions of Hillary after a first meeting in London are typically unstated. Also typically, both Hillary and Lowe were open with their assessment of Fuchs. Hillary described a very serious and perhaps humourless approach . . . more dogged than Hillary could ever be . . . a quite abrupt manner . . . 'Even at this first meeting I felt that Fuchs and I had little in common.' That sounded right.

Later, when Hillary joined Fuchs on a voyage to the aptly named Shackleton Base on the Weddell Sea in the 900-ton *Theron*, that feeling was strengthened. Hillary would say later that he felt an outsider even though they shared the same cabin.

George Lowe later described climbing with Hunt and Hillary as having an element, naive but enjoyable, of a boyish pirate game. 'Crossing Antarctica under Bunny's leadership it was as if it were controlled by the

A present for Peter — Sir Edmund brings back a midget pair of Norwegian skis after his preliminary visit to the Antarctic with Vivian Fuchs in the icebreaker *Theron*.
NEWS MEDIA

disciplined direction of a fine headmaster.' In another typically schoolteacherish and thoroughly Lowe fashion, he described notification of a Fuchs party briefing being relayed as 'Matron wants to see everyone before school!'

Again, for apparently no obvious reason, Fuchs banned music at Shackleton Base except at weekends. Clearly, Fuchs and the man he would depend on so much for success in the venture were totally different people, in almost everything except their overpowering obsession to succeed.

Their roles were as clearly defined as their characters. Fuchs' journey from the Weddell Sea would necessarily involve a series of supply depots beyond the Pole to sustain his party as it moved towards the Ross Sea. A New Zealand party, led by Hillary, would carry out that task from the newly established New Zealand Scott Base using Ferguson tractors converted for snow and ice work.

That simple-sounding assignment was further boosted by a New Zealand Government decision to send a scientific party with the Hillary expedition as part of world research marking the International Geophysical Year. The Royal New Zealand Navy would provide the expedition vessel and RNZAF pilots and an Auster. A public appeal would help fund the venture — with Hillary's presence and name as a huge drawcard.

In the summer of 1955–56, Fuchs took an expedition south into the Weddell Sea to set up his Shackleton Base. Hillary, Bob Miller, the New Zealand expedition deputy leader, and Squadron Leader John Claydon sailed in the *Theron* with him. Very quickly, the ship had a glimpse on a much smaller scale of the problems that had afflicted Shackleton and forced him into his desperate journey of survival. The *Theron* was pinned for days in the pack ice before finally making an icefall and beginning to unload the base gear.

This was a learning time for Hillary and the New Zealanders, while

Moment of farewell — the polar adventurers prepare to sail from Bluff in the *Endeavour*. In the middle, holding his son, is the deputy leader of the New Zealand party, Bob Millar.

NZ HERALD

Under the wing of the expedition Beaver aircraft, Edmund Hillary checks the fittings from which the bright orange fuel drum will be 'bombed' on to the snow for pick-up by the vehicles.
TRANS-ANTARCTIC EXPEDITION

Fuchs applied the experience he already had from his years in the region. Inevitably, Hillary learned his lessons quickly and was intent on putting them into practice. Just as inevitably, the differences in their approaches and their personalities became obvious; those differences would ultimately lead to world headlines.

George Lowe — as someone who knew Hillary better than anyone and who was shrewdly assessing Fuchs, his new leader — saw the trends. At this stage, it was a matter of things rather than tactics. Hillary believed changes were needed in detail which had already been fixed — at least in the mind of Fuchs. Hillary modified the design of the party's huts to allow every man his own tiny but personalised bunk space. Fuchs favoured a central bunk room. Hillary was astounded at the stoicism/masochism/lack of thought of the British who failed to build a heating system into their Sno-Cat vehicles. The two men thought and acted very differently. It showed immediately and more markedly as the expedition wore on.

Both groups in the expedition — Fuchs on the Weddell Sea, Hillary on the Ross Sea coast — arrived on the same day, 4 January 1957. Fuchs marked the occasion with an untypical little flash of humour. His signal: 'Snap!'

The completion of Scott Base was the first priority while, at the same time, field parties and aerial surveys familiarised themselves with the area around them, establishing a likely route from the base to the polar plateau. In addition, Hillary took advantage of a United States Globemaster invitation flight to the Pole to check the route he would later follow on the ground. Already the Ferguson tractors, unlikely vehicles to drive into history, were proving their worth — unspectacular, maybe, but also tough and reliable. In the polar region, no one could ask for more. As well, the dog teams, that would act as scout and support units on the major polar journey and many others, were revelling in their work. Among those setting the sled pace was Harry Ayres, so long ago Hillary's mentor in his first major climbs. In a life of Hillary, the same significant names keep recurring.

More than the dogs were straining at their harness. In what was nominally a trial of the Ferguson tractors but was so obviously Ed Hillary being bored in one place and counting down to the main thrust of the expedition, Hillary decided on a trip to Cape Crozier, about fifty miles from Scott Base. They would follow the route blazed into history by Wilson, Bowers and Cherry-Garrard more than forty years before: *The Worst Journey in the World*, as Cherry-Garrard described it in his classic book. A copy of that book went with Peter Mulgrew, Jim Bates, Murray Ellis and Ed Hillary.

This journey represented a new link with old history as they located and explored the stone hut the Wilson party had built. There were traces too of the mission that had drawn Wilson to the area, winter rookery of the emperor penguin. Wilson had believed that a study of the embryo of the penguin might reveal the origin of all birds by establishing a link between reptilian scales and feathers. A fine artist and an ornithologist, Wilson was prepared to go through a version of hell on earth to test his theory.

When the Wilson party slogged their way back through terrible weather — the conditions that would ultimately bring Wilson to his death with Scott — they carried with them three emperor penguin eggs. Behind them, at the hut, they left the dried carcasses of some of the penguins they

One of the expedition dog teams in full cry.
TRANS-ANTARCTIC EXPEDITION

The tractor train under way with its load of fuel and supplies.
TRANS-ANTARCTIC EXPEDITION

had used in a bid to get warmth from their blubber stove. The eggs went on to British ornithologists, while beautifully detailed colour plates of the penguins exist as reminders of a dead man's dream.

There were flashes of the wind that had flayed the three Englishmen, but fortunately none of the extreme cold that all but overwhelmed them. On the pages which Hillary and new teammate Peter Mulgrew studied in their tent on their version of the journey was this description by Cherry-Garrard: 'Such extremity of suffering cannot be measured. Madness or death may give relief. But this I know: We on this journey were already beginning to think of death as a friend.'

In that icy rock-walled hut, the three men had lived for four weeks, praying, singing hymns, supporting each other until they were able to free themselves from the storm and stumble back to Cape Evans with those three eggs but no basis for the theory which had drawn them there. When the three New Zealanders revisited the desolate Cape Crozier site decades later, they found traces of those days long before — the dried penguins, a collector's eyeglass amongst them, plus four of Wilson's pencils, and fragments of the torn canvas that had once acted as a roof on a blizzard-swept hut.

Hillary's tractors covered the nearly fifty miles back to Scott Base in fourteen hours. The long walk back had lasted six terrible days for Wilson and his men.

CHAPTER SEVEN

Coming — Ready or Not!

During the long polar winter that followed — the darkness settled in when they lost the sun in mid-April — the Hillary party at Scott Base began preparations for the long journey to follow, preparing the three Ferguson tractors and building a new vehicle based on the experience of the journey to Cape Crozier. Labelled 'the caboose', it was a caravan on skis with bunks, cooking and radio facilities as a mobile rest haven for what lay ahead.

This was another new stage in the life of Ed Hillary. Already a skilled and courageous explorer, he was increasingly in need of new leadership skills. This was a long way both geographically and personally from those small parties of individuals in the far-away Himalayas, doing their own thing, making their own arrangements, answerable only to themselves as a group.

The New Zealand operation was administered by a group known as the Ross Sea Committee in Wellington, headed by retired cabinet minister Charles Bowden. Although Hillary at first found the committee easy to deal with, once he was on the ice the relationship was sometimes strained. He felt, understandably, that they did not really grasp the issues involved in either the day-to-day running of the organisation or the long-term aspects either.

Hillary later summed up: 'Inevitably, I came to regard the committee as one of my heavier burdens — while the committee developed a healthy suspicion of my actions which resulted in constant efforts (somewhat in vain, I fear) to exert a moderating brake.' As before and after, Hillary resolved the potential problem by simply pressing on in his own way. Already an able judge of men and a field leader, Hillary was doing a crash course in the *Yes, Minister* processes of distant bureaucracies.

Across the ice at Shackleton Base, Vivian Fuchs was going ahead with his plans in a station where everything ran to a programme Whitehall would have found absolutely acceptable and totally unsurprising. Fuchs was leader, fullstop. He made the decisions and the party carried them out. George Lowe, who spent the best part of three years with him in the course of the Trans-Antarctic enterprise, talks about Fuchs with absolute respect and admiration but without the very real affection so obvious in his relationships with both Hillary and Hunt. As leader, Fuchs did not seek nor want the opinions or the judgements of others. He simply passed on to others the decisions he had made.

The difference in the two styles was probably best illustrated once the vehicles started rolling across the ice. Hillary's party drove theirs sometimes right around the clock using a shift system, with each man taking his turn at the head of the column while others followed or rested. Bunny Fuchs used what George Lowe called 'a Napoleonic type of command'. He drove at the front from start to finish. 'When he tired, we halted. When he was refreshed, we moved on.'

That's how it was when the two vehicle parties headed out from either side of the ice continent in October and November of 1957 on the great journey. The Hillary party was moving up the Skelton Glacier with tractors and dog teams to lay supply depots between the Ross Sea coast and the Pole. The Fuchs group of eight vehicles, sledges and dog teams — twelve men, in all — was to make a crossing of more than two thousand miles of unexplored snow and ice.

Looking back nearly forty years later, it seems obvious that the linked movement schedules of the two parties very quickly slipped out of mesh. Hillary's party was already five weeks and 350 miles on their journey, with the Plateau Depot already laid, when Fuchs radioed them with news that he was only then about to leave his South Ice base on the main journey — a fortnight behind schedule. Events were rapidly moving towards a very public difference of opinion.

Hillary said later that the new timetable Fuchs had set himself 'devastated' him, carrying with it what he believed was the risk of another winter on the ice for both parties. The time difference did not lessen as the Hillary party slogged on across sometimes crevassed ice on the polar plateau. Peter Mulgrew had been forced to withdraw temporarily to Scott Base with broken ribs. Ed Hillary battled through a bout of high fever.

By 6 December, the Hillary party had stocked up Depot 480, as it was called, and moved on dragging sledges with a further eleven tons of supplies for Depot 700, the official turnaround point for the party, only

Sir Edmund Hillary shouts his goodbyes as his tractor train heads out from Scott Base to begin the long crossing. Even then, he had plans to get to the South Pole if possible.
TRANS-ANTARCTIC EXPEDITION

A camp and dog teams in blizzard conditions.
TRANS-ANTARCTIC EXPEDITION

The Polar Plateau camp, 280 miles out from Scott Base.
TRANS-ANTARCTIC EXPEDITION

The pressures of the journey and its politics forgotten, Edmund Hillary greets Vivian Fuchs at the South Pole.
ROYAL GEOGRAPHICAL SOCIETY

Following pages: The British party gets a blasting from high winds and snow.
ROYAL GEOGRAPHICAL SOCIETY

The British dog teams take a well-earned rest at the South Pole. Behind them are the vehicles of the crossing party.

Sno-cat *Rock 'n' Roll* with Transantarctic expedition leader Vivian Fuchs and David Stratton aboard comes to grief above a crevasse. They escaped, Fuchs crawling over the track as it hung in space.

ROYAL GEOGRAPHICAL SOCIETY

500 miles from the Pole.

But already a different scenario was building up. The exact sequence of events and decisions which would ultimately take Hillary to the Pole is difficult to trace now. But one thing is clear: it was not at all a plan that came suddenly out of the polar air. He had talked openly about it to close associates before he left New Zealand. Official documents show that there had been some consideration of the possibility as early as March — nine months before it became a real issue — when Hillary had first floated the plan. He had a call then at Scott Base from Charles Bowden expressing the Ross Sea Committee's concern at the possibility that Hillary might take a party to the Pole, 'spoiling Bunny's effort'.

Soon after, in a confidential report to the committee, Hillary said: 'After discussion with Dr Fuchs and Mr Bowden, I have decided to shelve any possible plans to get a New Zealand party to the South Pole.' Just when and if the discussion with Fuchs took place is not clear, but what does seem plain is that any Pole proposal had been scrapped.

All that makes strange reading of a much later message to Hillary from the committee when he was only 250 miles from the crucial Depot 700, talking of

> . . . greatly increased public interest in expedition . . . committee interested in your prospects reaching Pole and whether you have considered this. If you are prepared to go for the Pole committee will give you every encouragement and full support following formal approval from London. If you intend to proceed Helm [the committee secretary Arthur Helm] requests you seek committee approval for the venture following which they will get O.K. from London.

Hillary quotes himself as 'astonished, to put it mildly'. For good reason. If his dreams of the Pole had ever really gone away — and, knowing the man, that seems very doubtful — then this would surely have refired them.

An already strange situation was complicated even further by what Hillary later described as a misunderstanding of a crucial part of that message. He says he took from the rough copy he was handed that London approval had already been gained. He says he misread 'following formal approval from London' as meaning that permission had been granted. Given the full text, it seems surprising that he would, but that is his recall. He began doing his sums to see if a dash from Depot 700 to the Pole was possible.

Meanwhile, back in Wellington, the Ross Sea Committee did one more of its U-turns. It directed that Depot 700 'must not be left unmanned since experience proves this may make it impossible to find under certain climatic conditions'. The message suggested that any change in plan — like going to the Pole, presumably — should await a meeting with Fuchs 'at Depot 700 or such other point as necessity requires or as requested by Fuchs'.

Hillary was very angry, and unconvinced, as it turned out. He headed off under a great head of steam for Depot 700, where you would take from this message he had been directed to stay put. The tractor party survived more crevasse problems to arrive there on 15 December. But stay it did not. The latest fix on Fuchs at that stage gave him a suggested arrival time at the South Pole between Christmas and New Year.

That Pole idea had certainly not gone away. Hillary went back to his figures. On average consumption so far from Scott Base, they had just enough fuel to make the Pole but with virtually nothing in reserve. As far as time was concerned, he reasoned

> . . . there was little likelihood of us getting to the Pole by New Year though it shouldn't be too many days after . . . as long as we were within a couple of days' march of the Pole when Bunny arrived I thought he would be agreeable to us carrying on and finishing the journey with the Fergusons. And, in any case, the further out we proved the route through the crevasse areas, the quicker we would all get back to Scott Base and safely home.

A very neat piece of rationalisation. Off went a message to Fuchs telling him of the altered plan. Hillary was, he said, out 'proving the route another 200 miles and then, if the going proves easy, doing a trip to the Pole . . . will scrub southward jaunt if vehicles and fuel can be used in any way to expedite your safe crossing either by a further depot or anything else you suggest . . .'

Incoming signals included one from the now decidedly edgy Ross Sea Committee chairman Charles Bowden drawing attention to earlier messages of 5 and 17 December and ending 'meanwhile you should not proceed beyond Depot 700' (already twenty-seven miles behind them). Hillary's response was that he seemed to have mislaid the signals. Real Lord Nelson 'I see no signal' stuff this.

Fuchs caught up with Hillary by radio when the New Zealand party was 200 miles clear of Depot 700 — their nominal destination and the point the last Bowden signal had specified they should not leave. By now, even the unflappable Fuchs seemed somewhat rattled by events. He discussed the available fuel at various depots, including 700, and took up the earlier Hillary suggestion that they might lay a further depot to protect the party against possible fuel shortage. Within that message was a significant punchline.

> Am in difficult position of feeling I must accept your offer to clear

Point of decision — Edmund Hillary, Murray Ellis, Jim Bates, Peter Mulgrew and Derek Wright at Depot 700, where Hillary opted for a run to the Pole.
TRANS-ANTARCTIC EXPEDITION

> present crevasse area then establish additional fuel depot at appropriate position from D700 thus abandoning your idea of reaching Pole. Know this be great disappointment to you and your companions but the additional depot will enormously strengthen the position of the crossing party which cannot afford to deviate from the direct route . . .

Once again, Fuchs was telling Hillary not to go to the Pole. It was a nudge which Hillary did not feel, a wink he did not see. He saw it simply — and perhaps accurately — as an excuse by Fuchs to stop him going to the Pole. He signalled back to Fuchs that his message had arrived too late. His own fuel situation was such that he could now either return to Depot 700 or go on to the Pole, but the party did not have food or fuel to await Fuchs' arrival. 'We will wait for you at the Pole' (at the United States Operation Deepfreeze base there).

Interestingly, as well as Bowden and Fuchs, others had misgivings about the tractor sprint for the Pole. According to Hillary in *Nothing Venture, Nothing Win* both Murray Ellis and Jim Bates of the New Zealand tractor party expressed concern to him about the risks if the weather deteriorated on the 500-mile journey to the Pole. They had misgivings that they might find themselves out of fuel and beyond the effective range of their Beaver support aircraft. Hillary reports himself as pretty grumpy — 'I seemed to be in a minority as far as my plans were concerned.' He was right: Bowden, Fuchs and now his own party companions had misgivings.

> Should I just call it a day like everyone wanted me to? . . . I was sure that my plans were feasible enough even if our margins were a little slim. In a pinch, we would just have to drop one vehicle — or two. We would be carrying manhauling sledges and in an emergency we could complete the journey on foot — as Captain Scott had done forty-six years before.
>
> I was certainly getting a little obsessed about the whole business and probably the only action I never considered seriously was to turn back.
>
> I even worked out the logistics for heading south myself with one tractor, although I was pretty sure that Peter Mulgrew would go with me whatever happened.

Peter Mulgrew — he would have gone with Hillary to the Pole if no one else would.
TRANS-ANTARCTIC EXPEDITION

He was right. Like Hillary, Mulgrew could not have resisted the challenge.

Hillary's own word 'obsessed' is an interesting description. So is the reference to Scott. And the suggestion he might go solo. Certainly, it is clear that nothing would stop him. And neither it did. Not the weather and the effect of their 11,000-foot altitude that produced biting cold and tiredness. Not Bowden nor Fuchs. Hillary was going to the Pole! Just as he had planned months before. First there since Scott.

With that decision made, Ed Hillary made another sincere but highly controversial one. He sent off a message to Fuchs which, when combined with his apparently impromptu journey to the Pole, caused real controversy. Hillary's message was totally unexpected, and yet in his mind completely justified. He talked later of 'a conviction that the crossing party was not motivated by any particular sense of urgency . . . there was a very real danger of the expedition getting into serious trouble unless it developed much greater impetus'. Having decided to head off Fuchs to the Pole — because however you justified the decision, that was the effect of it — he now tried to convince the Englishman that he should temporarily abandon his great epic.

> Dear Bunny,
>
> I am very concerned about the serious delays in your plans. It is about 1250 miles from the Pole to Scott Base, much of it travelling from D700 north being somewhat slow and laborious with rough, hard sastrugi [fluted, wind-carved ridges]. Leaving the Pole late in January you will head into increasingly bad weather and winter temperatures, plus vehicles which are showing signs of strain.
>
> Both my mechanics regard such a late journey as an unjustifiable risk and are not prepared to wait and travel with your party. I agree with this view and think you should seriously consider splitting your journey over two years. You will probably have a major journey in front of you to reach the Pole. Why not winter your vehicles at the Pole, fly out to Scott Base with American aircraft, return to civilization for the winter and then fly back to the Pole Station next November to complete your journey . . .
>
> Personally, I feel the need for a break from the plateau after nearly four months of tractor travel and there's a lot to do. I prefer not to wait at the Polar Station but will get evacuated to Scott Base as soon as possible. If you decided to continue on from the Pole I'll join you at D700.
>
> Sorry to strike this sombre note but it would be most unfortunate if the sterling work you've put in to make your route through to South Ice and the Pole should all be wasted by the party foundering

> somewhere on the 1250 miles to Scott Base. I will go ahead with the stocking of D700 and I will leave at the Pole Station full details plus maps of the route from Scott Base to the Pole.
>
> Ed Hillary

The Fuchs response was calm and measured. It was also very straight to the point.

> Appreciate your concern but there can be no question of abandoning journey at this stage. Innumerable reasons why it is impractical to remount the expedition after wintering outside Antarctica . . . I understand your mechanics' reluctance to undertake further travel and in view of your opinion that late season travel is an unjustifiable risk I do not feel able to ask you to join us at D700 in spite of your valuable local knowledge. We will therefore have to wend our way using the traverse you leave at the Pole . . .

Hillary says he did not take seriously the Fuchs suggestion that the Englishman would press on alone, which probably tells something of how little he really knew Vivian Fuchs. Nothing would stop *him* either — certainly not some unsolicited and, he believed, badly based advice from Hillary.

Hillary then sent a copy of his message to the London management committee 'suggesting they use their influence on Bunny to re-examine his position'.

That was when problems really began. Somewhere along the way — probably in Wellington — the message leaked to the Press and the combination of the polar trip and this suggestion that Fuchs go home became an international issue or, as Fuchs would later describe it in his official account of the expedition, a *cause célèbre*. It was certainly that. Then, and later, Ed Hillary waved both issues away as largely of the media's making. Storm in a teacup, was one of his descriptions. It was clearly more than that.

It could have been a major disaster for the expedition but for the personalities of the two key figures. As it was, Hillary went to the Pole and reached the point of his ambition on 4 January 1958, with twenty gallons of fuel left — enough for only twelve more miles. Fuchs arrived at the Pole on 20 January to be met by Hillary — who had flown out to Scott Base and back to be there — on an occasion which did both men credit: there was no hint that there had been any problem, no sign of feeling or tension. From George Lowe came one of those typical responses as he climbed out of his Sno-Cat and extended a gloved hand to his old mate: 'Sir Edmund Hillary, I presume.' It was indeed, with mail that included an income tax demand for George.

The face of triumph — Hillary at the Pole.
THE PRESS

Nor was there any hint of problem when Hillary joined the Fuchs cavalcade at Depot 700 to lead them through what to him was now familiar territory back to Scott Base, sometimes even following the track marks made months before by his trusty Ferguson tractors, arriving on 2 March. Vivian Fuchs, quickly made Sir Vivian, had achieved his own personal goal. Planning to cross in a hundred days, he had reached that far coast in ninety-nine. It was a triumph for Fuchs, his planning and his will.

It was also an important occasion for New Zealand's involvement in both that expedition and the Antarctic. Apart from the spectacular events on tractor and dog sled, the party had set up Scott Base, explored, surveyed and made geological surveys of thousands of miles of previously unknown territory. All these things were enduring assets for world research.

How the world saw the first polar meeting between Hillary and Fuchs — a wire photo of the historic moment.
TRANS-ANTARCTIC EXPEDITION

Hillary and Fuchs at the Pole — no cross words, only smiles.
NEWS MEDIA

Shackleton and Scott had been at daggers drawn for many of the final years of their polar involvement nearly fifty years before. Scott, in particular, believed that Shackleton had exploited his Antarctic base ungraciously. They never spoke — even when they shared the same lecture platform. The potential for differences between Fuchs and Hillary was much greater. But, if such rifts were there, they never showed. Media observers, keen to identify and record what could have been more world-wide headline material, failed to detect even a hint.

Yet, no one could have blamed the two leaders for tension. On the one hand, Fuchs treated the whole race-to-the-Pole affair as if it never happened. Hillary waved it away as some bizarre media creation. Even the normally politically uninvolved George Lowe felt the need to provide a non-confrontational account of the whole incident, which was, not surprisingly, a clear defence of Ed.

> One thing I know better than most people, from my experience of Hillary on our expeditions: there was no malice in Ed's character and actions. He does most things for the intrinsic joy of them — and getting to the Pole in three commonplace tractors, having finished his assignment with time to spare, was his method of brightening the bleak cold weeks of 'hanging around'.

> Ed told me: 'Quite simply, our feeling was that there was no point in waiting. We'd established the depot, we'd brought in the fuel and the food. What was the use of hanging around?'
>
> Moreover, Ed and his party were living in tents. There was no hut at Depot 700. So the prospect of waiting several months on the polar plateau with nothing to do had no appeal for anyone.

Lowe believes that Hillary's stated intention of clearing crevasse areas on Fuchs' route was a prime reason for his move. And he describes the controversial advice to delay the second half of the crossing, from the Pole to the coast, as 'acting with a sense of responsibility'.

> It seemed to me, when the shouting eventually died down, that both men had behaved in a logical as well as a characteristic fashion.
>
> The restless Ed Hillary whose 'jaunt' to the Pole had no selfish, glory-seeking intentions — and who made our speedy exodus possible — just did not realise the potential of our Sno-Cats in the last stages of the crossing. In different circumstances, his counsel could have been right . . . it was proved wrong.
>
> The enigmatic Bunny Fuchs, whose inner motives baffled us, had with dogged persistence got us across the bottom of the world as planned. He could have been wrong . . . he was proved right.

It's a typically neat and non-judgemental assessment from a thoroughly nice guy. But I'm not at all sure that it is borne out by Ed's own explanation of what happened and why.

> What did we achieve by our southern journey? We had located the crevasse areas and established the route and we had been the first vehicle party to travel overland to the South Pole — that was something I suppose. But we had produced no scientific data about the ice and little information about its properties. We showed that if you were enthusiastic enough and had good mechanics you could get a farm tractor to the South Pole — which doesn't sound much to risk your life for. The Press had a field day on the pros and cons of our journey but for me the decision had been reasonably straightforward.
>
> I would have despised myself if I hadn't continued — it was as simple as that — I just had to go.

That's it in five words. He had to go. There is no need for great rationalisation about crevasses and the rest. That ambition apparently abandoned nearly a year before simply had not gone away. 'I just had to go.'

Also there is a key fact. 'We had been the first vehicle party to travel overland to the Pole.' What is not referred to is that this 'first vehicle to the Pole' feat was also implicit in the Fuchs planning of which Hillary was a key part. Fuchs believed he would do that, from the other coast. For one good reason or another, Hillary topped Fuchs off.

Very early in my research for this book, I remember being puzzled by a response to one question. It was put to someone who was very close to Hillary — their identity doesn't matter. When I sought their response on the significant features of Hillary's character, a long, long pause followed. The ultimate reply, among others: 'Well, he was ruthless . . .'

I felt I had misunderstood or that the choice of word was not quite accurate. 'By ruthless, do you mean single-minded?'

The escape route was not taken. Another long pause. 'I said what I meant.' The voice dropped away. 'He showed that in Antarctica.'

Months later, many interviews and many thousands of words later, I felt I knew a little better why that word had been used. I went back and re-read part of Peter Mulgrew's assessment of Ed Hillary, quoted as the foreword to this book, written by a man who, like George Lowe, knew Hillary better than any of his other contemporaries. In a reference written nine years after the Pole journey, Mulgrew said: '. . . yet he can appear surprisingly thoughtless, even careless of the feelings of others.' Vivian Fuchs could not have said it better — but, significantly, and to his endless credit, he chose not to.

Douglas McKenzie of the *Christchurch Star*, as the New Zealand Press Association correspondent on the ice, was close to the events and the people. Like my other observer, he too uses that unexpected description assessing Ed Hillary in the book he wrote later, *Opposite Poles*:

> Hillary was principally concerned about Hillary. Behind his easy-going manner he held that thread of ruthlessness which must be possessed in some degree by all successful leaders. With a casualness which was startling to those who met it, he was willing to place members of his party temporarily on the side-line if this became necessary for his major purpose.

Of the Fuchs-Hillary meeting at the Pole, McKenzie wrote:

> . . . the two men had a history of indifference to one another. They were two men who had been thrown together by force of circumstance and they were apparently without common ground for social intercourse. Their politeness towards each other was unfailing, as between strangers. Whether Fuchs had distrust for Hillary, and Hillary distaste for Fuchs was never revealed, nor did anyone expect it to be. The men were at the Pole together for about ten hours . . . but apart from formal sessions together, Fuchs and Hillary took their pleasures separately.

Mission accomplished — the newly honoured Sir Vivian Fuchs and Sir Edmund Hillary in a drive through Wellington streets to a civic reception.
NZ HERALD

Hillary talked about being obsessed with the Pole. He had to go. It was, in some ways, rather the equivalent of this: Imagine a New Zealand expedition committed to a traverse of Everest by Hillary, up the north face to the summit and down the south. Imagine that a party of British climbers had undertaken to prepare camps and cut steps to speed up the final descent. Having done so ahead of schedule, they then made the ascent themselves — the first ascent for forty years. Their explanation: 'We just had to go.'

I doubt whether Hillary would have been as accepting and stoical in his reaction. I believe that word 'bastard' from his first climb may have reappeared in a different context. Later, Hillary would say by way of explanation, or justification perhaps: 'I didn't realise just how anxious Bunny was to be the first to the Pole with vehicles.' Really? He only had to hold a mirror to his own feelings to have some guide to Fuchs' ambition in this matter.

For his part, Vivian Fuchs was as intent as Hillary in playing down the whole affair — and still is. He says that the Pole had never figured in his early planning, that since there was no station there originally his intention was to pass some miles from the actual Pole. (It seems unlikely, but that is what he says.)

Asked then why he changed his mind after the Americans put their base there, he replied: 'Because it would be extremely rude. "Why is this damned stuck-up Limey ignoring us?" So, one went in. We took from them, in the matter of supplies, one pot of jam. That was all.' The almost imperial belief in things British meant that later, when he discovered that Hillary had left some American fuel at one depot, he left it there. He had pledged that he would use only British.

Decades on, after the furore over Hillary and the published criticism of Fuchs himself by the New Zealander, how does Fuchs react? With complete equanimity and charm. Their relationship: 'We are perfectly good friends.' That message about temporarily ending the crossing: 'It was silly. And as I was in overall charge of the whole thing, I just said "can't do that" and went on. But it leaked. It became a *cause célèbre*. It was silly nonsense really.'

Vivian Fuchs is a great man with dogs. He even took a lead husky back to retire at his English home. He talked to one of my favourite journalists, Terry Coleman of the *Guardian*, about dogs and how to handle them. He explained that if you have trouble with a dog you take him out and put him next to the king dog and there is no more trouble, or you pair a troublesome dog with a bitch and there is no more trouble. One wonders what pairing he would have liked for a certain mountaineer at crucial stages of their relationship. But there is still no criticism.

> Hillary? He's a tough customer, who's a splendid mountaineer. I regard mountaineers as runners . . . they go there to the bottom of a mountain and they're up and down in two months. Well, I go and I expect to be two years, a long endurance. When the weather goes wrong, I'll sit. I'm not going to go back. And that way I will beat nature's problems. But it's different. That's the cross-country runner as opposed to the 100 metres. I mean, I think this is a reasonable comparison.

So, at the end of it all, after the celebration at Scott Base when Sir Edmund proposed a toast to the newly knighted Sir Vivian — 'errant knight and knight errant' Douglas McKenzie later archly described them — after the open car drives with Hillary in Wellington, the reunion with his wife, who was a guest of Louise Hillary, and the journey home to a British hero's welcome, the cross-country runner from the ice became Director of the British Antarctic Survey. He wrote a book about the crossing, which includes biographical paragraphs on everyone but Vivian Fuchs, and another about the history of the survey, which covers his trans-polar feat in a footnote buried in the middle pages.

George Lowe packed up his gear to leave Antarctica, including the battered copy of *War and Peace* he had read to fill the many long hours on the crossing. He had seen it first on Everest, where party member Michael Westmacott loaned it to John Hunt. Charles Evans had packed it with his paint box when he led the party which climbed Kanchenjunga and it had passed on to Lowe. Lowe still has it. The title page with dedications telling of its history has been mysteriously removed, but it still holds hints of pressed flowers from the Himalayan valleys, tea stains and a vague smell that might be pemmican.

Ed Hillary and newly bonded kindred-spirit Peter Mulgrew flew home to wives and families wondering perhaps where their adventurous ambitions might lead them next.

CHAPTER EIGHT

The Great Yeti Hunt

Amidst all the drama of his life and death, one memory of Peter Mulgrew remains vivid. We were standing around in a group making small talk at an Auckland business cocktail party, glasses in hand, tending to shout over the noise of those around us. Someone neither of us knew joined the group. Self-consciously, he introduced the subject of squash and babbled on for some minutes until the conversation lapsed again. He turned to Peter Mulgrew, desperate to pick up the verbal momentum. 'Do you play squash at all?' he asked. I felt myself stiffen. Peter Mulgrew looked into his empty glass and paused for a moment. 'Not much now,' he said.

I have often wondered when that unfortunate stranger realised that, as Mulgrew left a little later to refill his glass, he was walking on artificial legs. He had spared his questioner embarrassment.

Ed Hillary's memory was as typical. He recalled meeting Peter Mulgrew for the first time in interviews to select the final party for the polar support team. Both finalists were navy officers with excellent background in communications — a priority. But one added 'sir' to each sentence. Mulgrew didn't. A small thing, but it helped gain the place for Mulgrew.

There was, of course, much more to the man than an absence of 'sirs'. He had tramped and enjoyed skiing since his childhood, and even after joining the navy he had his gear and skis with him. No foreign mountain anywhere near a port was safe from him. This combination of abilities took him with the Hillary party to the Antarctic. Later, Mulgrew would talk about 'trudging disconsolately' towards the South Pole. This throwaway description doesn't do him justice.

Once on the ice, Ed Hillary quickly recognised a kindred spirit in Mulgrew, a man who was prepared to give 110 per cent, fiercely com-

Peter Mulgrew and his wife June after his investiture with the Polar Medal from the Queen at Buckingham Palace in 1959. The previous year, he received the BEM for his work at the Pole.

petitive, the one man Hillary believed would have gone to the Pole with him, if necessary manhauling a sledge. His whole life before and after the Pole stamped him as just one more of the exceptional people whom Hillary's life has drawn around him: Hunt, Tenzing, Lowe, Fuchs and certainly Peter Mulgrew. Theirs was one of the great friendships of both their lives.

There was an absolute inevitability about Hillary's invitation for Mulgrew to join him on his great new expedition and that Mulgrew should accept immediately. According to Mulgrew, the original Hillary intention was to attempt Everest again, this time from the north through Chinese-occupied Tibet. Hillary does not discuss this option in his accounts of the expedition, but since the two men were so close there seems no reason to doubt the Mulgrew report of the plan and its abandonment after the Chinese apparently refused access.

Instead, Hillary decided to attempt Makalu, the mountain which had so held his gaze in those minutes on the summit of Everest and whose profile had dominated Dan Bryant's study the first day I heard the name of Hillary. The French had already climbed the peak using oxygen. The Hillary party would attempt it without.

This, though, would be an expedition with two major differences. As if unwittingly responding to the then unspoken Fuchs comparison of climbers and their short attention span, Hillary planned a long presence in the mountains. His aim was to see whether a prolonged period of life in the high, rarified air would condition climbers at least to a degree where the 27,790-foot summit of Makalu would be reached without oxygen. Parties had reached that altitude on the easier slopes of Everest; Hillary believed the same might be possible on the testing upper reaches of Makalu.

Before putting that theory to the test, Hillary wanted to seek proof of that accepted folklore of the Himalayas — the existence of the Yeti, the Abominable Snowman. This venture alone would have won the backing of *World Book Encyclopedia*, which sponsored the expedition of what Hillary described as his dream blend of climbing and science.

Others beside Mulgrew were drawn to go with him. Years before, after the Shipton reconnaissance in 1952, Hillary and Lowe had spoken about their experiences to an audience in a small church hall in Auckland. Back in the crowd, and rather spell-bound by it all, was a young schoolboy called Michael Gill. That talk was a revelation — 'They were great performers' — and one of the spurs that took the young Gill into high country as soon and as often as possible afterwards. Later, he would talk of his teenage climbs with the fervour and eloquence of a lover. Going to medical school in Dunedin only brought the challenge of the mountains that much closer.

Back in Auckland on university holiday in 1959, a paragraph in my old newspaper the *Auckland Star* lit a long-running fuse. Sir Edmund Hillary, it said, was looking for two young climbers to go with his party. The twenty-two-year-old Mike Gill, with a Bachelor of Medical Science degree in physiology, thought he knew just the man to fill one of those gaps. He wrote a letter including a wonderful, whimsical description: 'I have an ape-like build peculiarly suited to climbing.' The candidate didn't leave anything to chance. He walked down the road to the Hillary home in Remuera and pushed his letter into the letterbox.

Next afternoon, he had a summons: 'a voice barked out of the earpiece at me', which proves that Ed's sometimes gruffish telephone style is not a late-in-life development. When he knocked on the Hillary front door within the hour, Louise Hillary greeted him with smiling interest. 'We've been dying to know what this ape-like person looked like,' she said.

Enthralled by Hillary's description of the plan, the young Gill then had to somehow survive the weeks between that interview and the final party choice. He filled it the best way he could — climbing. A month later, another phone conversation. 'Are you still interested in this expedition? Well, you may as well come along.' Ed Hillary always had a great economy of words.

The young Gill was suddenly into the unfamiliar life of major expedition preparation, sharing the packing with Hillary and Mulgrew. 'That was real fun. Ed always had a good sense of humour. He and Peter Mulgrew were great together, there was always a lot of banter going on. They sparked off each other all the time. It was the sort of thing which changed quite a lot later.' Disasters would change them both, but that was for the future. What Mike Gill could sense then was the excitement and anticipation, the pleasure of old companionship renewed.

The Great Yeti Hunt was an interesting combination of sceptical searchers — Hillary among them — who had little confidence that they would resolve the issue, unmask the myth, or reveal the truth of the legend, involved in a genuine scientific study. The fact was, the literature was there. The Shipton expedition to the area in 1951 had brought back from the Menlung Glacier photographs of large mysterious prints in the snow. The classic *The Snow Leopard* by Peter Matthiessen meticulously lists sightings of similar traces. He describes a young biologist from a field project in the Arun Valley producing a plaster cast of one footprint which he had found outside his tent six months before.

The Sherpas and Tibetans have no doubts. An abbot of Thyangboche monastery responded to a question from John Hunt in 1953 with a graphic account of a winter day a few years before when a Yeti had appeared just below the monastery where the expedition tents were. It had loped about in full view of the monks, sometimes on its hind legs, sometimes on all four, above five feet tall and covered in grey hair — a description Hunt was to hear often. The abbot even gave a demonstration of how the Yeti had scratched itself, had picked up snow and played with it, grunting now and then. After the monks sought to drive it away with a chorus of noise from conch shells and trumpets, it ambled off.

The Sherpa Tenzing also had a collection of Yeti sightings to recount, including how his father had met one face to face before Tenzing was born. It had confronted him on the Barun Glacier, like a big monkey or an ape with a pointed head, grey, a female who held its breasts in its hands when it ran. One feature Tenzing reported was that its hair grew

A normal boot print alongside the snow prints the Sherpas believe are those of a Yeti.
THE PRESS

'Yeti' tracks found by the Hillary expedition at an altitude of 18,000 feet.
THE PRESS

in two directions — upward above the waist and downward below. Tenzing's father had watched as the animal turned and climbed a steep mountain slope, making a high, shrill whistle.

Later, the elder Tenzing had a second sighting, when he was going cross-country to visit his son on his first Everest expedition in 1935. He was woken one morning by a sound outside his tent, a whistling sound again. When he raised the tent flap, he saw a creature coming down the Rongbuk Glacier.

Tenzing could quote dozens of such stories from all over the Himalayas — and frequently did. In Solu Khumbu, around the village of Targna, Yeti are supposed to have caused such damage to crops and buildings — sometimes trying to rebuild damaged sheds or replant crops after they had tampered with them — that villagers left out bowls of chang, the strong Sherpa beer, and Nepalese kukris, the dangerous curved bladed weapons of the area. When they came back in the morning, the reports say, they found the bodies of Yeti who had got drunk on the chang, fought and killed each other.

Tenzing reports that Julian Huxley, who once came into the area to research the mystery, told him he believed the Yeti sightings were actually bears. Tenzing's theory: 'I do not think my father was a liar and made stories up out of his head . . . certainly the tracks I have seen both on the Zemu Glacier in 1946 and near Everest in 1952 do not look like those of any familiar creatures. Though I cannot prove it, I am convinced that some such thing exists . . . an animal, not a man, that moves about mostly at night.' He tells how a porter with the Swiss expedition in 1952 had seen a creature with pointed head and heavy brown hair, which walked upright on its hind legs, made a hissing sound and ran away.

Peter Matthiessen recalls how one Sherpa told of his grandfather's accounts — how Yeti were once common in the Khumbu region but that the local people poisoned them off with treated barley to stop them raiding crops and that there had been dead Yeti everywhere.

The Hillary party of 1960–61 didn't want Yeti dead but alive and included two animal experts especially for that reason. Sponsored as it was by *World Book Encyclopedia*, the group had plenty of academic firepower. The Yeti department had drawn R. Marlin Perkins, director of the Lincoln Park Zoo in Chicago and a noted American television nature programme frontman with an armoury of spotting telescopes, tape recorders, cameras and trip wires.

Long after the party's return, and despite the effects of the events on Makalu, Peter Mulgrew could still raise a rich vein of whimsy about Marlin and his Yeti capture plans which involved the use of special 'Capshur' guns firing hypodermic needles filled with tranquilliser. The idea was to hit them with this hard stuff allowing ampule time — the pun is intentional

— for photography, examinations, measuring, paw printing, the lot, before a somewhat bemused Yeti was released back to retell its side of the story to its sceptical family and friends. As Mulgrew quoted one cynical newspaper editor: 'With nothing worse than a puzzled look in front and a dab of iodine behind.'

It never got that far. In fact, The Great Yeti Hunt fell a long way short of that. It did not find a live Yeti but it did detect a very healthy Yeti industry. Even Mike Gill, who, on his first Himalayan expedition could have been excused for a little wide-eyed naivety, detected that very quickly. Offered a 'genuine' Yeti skin by one enterprising merchant, he identified it instantly as the last relic of one very long-dead dog. Other 'genuine' skins were clearly bear, some of them the rare Tibetan blue variety.

Sir Edmund Hillary discusses planning for the party to the Makalu area with Dr Jim Nevison, an American climber on the expedition. Later, Nevison would come close to death with Peter Mulgrew on Makalu.
NEWS MEDIA

There were footprints like those sighted by Charles Evans and Geoff Harrow on the 1954 Hillary expedition. The 1960 team sighted a variety of large footprints on snow in the Ripimu Glacier area that Sherpas immediately and excitedly identified as Yeti. They were large and broad with clear toes. But, when they were followed into areas of shade, they shrank and became the obvious tracks of fox or wild dog. Experts believed that this effect of the heat of the sun in melting snow probably explained mysterious footprints which the Swiss party had found in 1952. Those prints were nearly a foot long and five inches wide.

Edmund Hillary with the skin of a Himalayan blue bear, which experts believe is a likely cause of some 'Yeti' sightings.

THE PRESS

Sir Edmund returns a Yeti 'scalp' to its homeland after exhaustive tests in the United States and Europe proved the relic to be a fake.

THE PRESS

Determined to resolve the issue of apparent Yeti scalps held as sacred objects in several places in the area, Hillary negotiated to take away a scalp and a skin for scientific study in Paris, London and Chicago. The finding: That the Yeti scalp from Khumjung, which had been escorted by a village elder on its long flight, was a fake made from an antelope; the skin was from a blue bear. A skeletal hand with tufts of hair attached from Pangboche monastery was shown to be human.

A journalist with the party, Desmond Doig, who could talk fluently to the people involved in their own language, was unable to substantiate one sighting. So, when Ed Hillary flew back from those laboratories so far away to return the sacred items he had been given in trust, some questions had been unanswered, some clues disproved, but amidst all the scepticism, there were still obvious out clauses.

Not the least of them, the lama who suggested that the party leave behind a camera when they moved on so he could photograph the Yeti who had deliberately kept out of sight while they were there and would come out of hiding when they left. And the Sherpa who told Peter Mulgrew: 'How can the sahibs expect to see a Yeti when they parade around the countryside like a herd of lumbering multi-coloured yaks?' As Desmond Doig said:

> I hope we are wrong about the Yeti. During our ten months in Nepal we never saw a bear of any kind. We saw surprisingly little animal life, nothing bigger than a wolf. If bears and snow leopards kept out of our way, an intelligent anthropoid certainly could. As I have said before, even our most convincing arguments are negative.
>
> Someone has yet to explain the commonly heard cry of the Yeti and keep looking for footprints in the snow that could have been made by an anthropoid or large bear and not a combination of brittle snow, a hot sun and a small animal's tracks.
>
> And someone has to discover conclusively what it is that Sherpas and Bhutanese and Sikkimese and North Indians are seeing when they claim to see the Yeti.

A good question.

CHAPTER NINE

The Silver Hut

If you look for a key figure in the adventure of the Silver Hut, then the search ends with a Dr L. G. C. E. Pugh, probably known to his parents as Griffiths, 'Griff' to his mates in the Himalayas and Antarctica. When the young Mike Gill got that call from Ed Hillary suggesting he join the party, one name in the brief conversation was that of 'Griff'. 'You might drop him a line,' Hillary suggested.

It was like offering a young maths student an introduction to Einstein. Pugh was a legend in his field. He originally trained in law but switched to medicine. He was a member of the British Olympic ski team in 1936, and trained ski troops in Lebanon on war service. After demobilisation, he joined the Division of Human Physiology of the British Medical Research Council, researching, among other things, adaptation to extreme conditions — heat, cold and high altitude.

That research took him to Everest as a scientific member of the Shipton expedition to reconnoitre Everest in 1952, where he met Ed Hillary for the first time. His work on and after that trip, involving the use of oxygen, his analysis of the need for a lengthy acclimatisation period and ample fluids were powerful factors in the success of the Hunt party on Everest the following year.

The popular belief is that problems at high altitude are caused by lack of oxygen. Surprisingly, it's not the oxygen that decreases at altitude, but the effect of lower barometric pressure. The percentage of oxygen remains a constant twenty-one per cent, but the pressure forcing oxygen into the lungs gets less, the higher the climb.

The body reacts by taking more air — and more oxygen — and increasing its red cells to carry the oxygen through the blood stream. Both the increased breathing rate and the red cell increase can produce medical complications. The lungs can lose three times as much water, causing dehydration, drying out the respiratory tract, raising the risk of fluid, pneumonia and other infections. Blood becomes stickier and more difficult to move around the body, lowering temperature and increasing the risk of frostbite.

Brain cells, the most sensitive to loss of oxygen, react, causing hallucinations and loss of memory, and an inability to make the right decisions. The heart is forced into greater activity because of sluggish vessels in the lungs, and enlarges. Climbers may lose consciousness and show signs of heart failure with pulmonary oedema (fluid in the lungs). All these facts become crucial in the happenings of Makalu.

But Hunt had also reported the symptoms as the earlier Everest party began to condition itself before the 1953 climb. Above 21,000 feet the party found physical effort and mental work much slower and sometimes painfully slow. Even routine tasks like setting up camp prove difficult. Everything takes longer and takes more out of climbers. The return from an ascent that has taken one or two days might require four and prove both mentally and physically exhausting.

It is difficult to sleep at altitude and parties frequently use sleeping pills. There are headaches and loss of appetite (Hunt sought to overcome this by having special packages of each climber's favourite food as an incentive to eat at the greater heights). Energy flags. Vomiting is common. Even some Sherpas react badly to altitude and are forced to retreat to lower levels to recover. It was a combination of all these factors that had forced the use of oxygen on Everest from the 1920s. Mallory and Irvine, for instance, used it in 1924.

Mike Gill tells graphically of his experience of the problems on an early part of the Yeti hunt. New in the Himalayas and exposed to the effects of high altitude for the first time, having 'raced' too quickly from 10,000 to 12,000 feet he found himself initially listless and then hit by severe headaches and sleep problems and affected by nausea. He turned blue, lost consciousness, was revived with oxygen and was unable to do more than walk very short distances for up to ten days. He was not fully fit for three weeks. This was classic high-altitude pulmonary oedema — 'I nearly died of it,' he told me thirty-three years later — and he would see more of it before the expedition was over!

Griff Pugh took his interest in these effects of cold and altitude with him to the Antarctic as part of the Fuchs party. Further discussions there with Ed Hillary — it was their third expedition together — prompted a plan for high-altitude study in the Himalayas. *World Book Encyclopedia* agreed to sponsor the project on a scale unimaginable anywhere but in the United States: $US200,000 is the figure Mike Gill quotes, big money in 1960. Mike Gill also quotes an American executive's account of a first conference with Hillary:

Here was this guy Hillary who'd climbed Everest and got a knighthood for it, the lot. We didn't know Ed too well at that stage — he was a nice guy, sure, but we knew he wouldn't be working in with an outfit like us for nothing. Probably, he'd have a bunch of lawyers and accountants with him — but we reckoned we could handle them.

Then in comes this tall guy, by himself, with his hair all over the place and carrying an old brief-case held together with string. Well, that threw us right from the start. And then when we came to the bit

Edmund Hillary gets a high-altitude haircut from George Lowe.
THE PRESS

where we asked how much he would like for himself he says: 'Well, on an expedition we don't usually take any money for ourselves.' We didn't know whether he meant it. For a bit we thought he might just be the coolest cat we'd ever met. Then, we began to feel sorry for him. We felt we had to help this guy — force him to take the money. Up till then I'd never been able to understand why he hadn't made a million bucks out of Everest.

I think Ed's the most honest guy I've ever met — I'm not sure he's not the only honest guy I've ever met.

A yak in a Sherpa village — the animal that serves for transport and provides long hair to be woven into clothing.
THE PRESS

On the strength perhaps of that meeting and that assessment, John Dienhard, PR manager for *World Book Encyclopedia*, went to the Himalayas with them. More importantly, perhaps, so did all that sponsorship money. They gave Hillary a new briefcase. Mike Gill picked up the old one and used it for years.

Out of all that came what by any standard were superb facilities for the Pugh study as part of a ten-month high-altitude winter-over by members of the Hillary party. They called it the Silver Hut. It was twenty-two feet long and ten wide, shaped like a tunnel with windows at either end and made of silver plywood. It housed eight bunks, a dining table and laboratory area and was wired to electric generators.

Among the aims was a study of the effect of sustained life at 19,000 feet and whether the months spent there would influence the ability of climbers to later attempt the assault on Makalu without oxygen.

Griff Pugh was the ringmaster of a prolonged series of activities in the months the party wintered over; Mike Gill was one of those in the Himalayas for ten months; others like Hillary, on his flight to Europe and the States with the Yeti relics; and Mulgrew, who also flew back to New Zealand with Hillary for some weeks, broke their time there. Later, there would be conjecture on whether this pattern was a factor in the events which overtook both of them.

The wintering party spent time each morning on a bicycle ergometer while samples of expired air and blood were taken and while leads carried their electrocardiography details. The group's work capacity in the first weeks was shown to be half that at sea-level, improving to about two-thirds towards the end of their winter spell. Pulse rates at maximum work levels were between 120 and 130, compared with sea-level rates of maybe 180 to 200, although they also improved. Mike Gill's blue colour on first exposure to altitude proved the shade of the season. They all turned blue under pressure of exertion. (A Sherpa who tried out on the cycle gear showed responses at a normal sea-level rate.) Mike Gill's colourful diagnosis of effects: 'Blood thickened with the rise in red cells and blackened by lack of oxygen oozed into a syringe from a vein like purple trickle.'

They all lost up to two stone in weight in their first months, and all the other classic symptoms were there: loss of energy, inability to sleep, lack of appetite and extreme exhaustion after any work. Even some of the most dedicated, like Mike Gill, had to retreat temporarily to lower altitude at times.

As well as work on the overall study, he had his own project designed by a Cambridge research unit, eager to test the effect of altitude on intellectual functions. For example, a list of ten letters would be read over a tape and the subject would then be asked to nominate which letter had been repeated twice. Comparisons seemed to show a slight deterioration at

19,000 feet compared with sea level, but physiologist Gill and his fellow subjects were not entirely convinced.

Inevitably, climbers did what climbers do. They attempted Amadablam, which was literally at their backdoor — the Silver Hut was on the Amadablam Col. It was a challenging peak, a monolith is one description, which had claimed the lives of two British climbers in 1959. Last seen high on the peak's north ridge, like Mallory on Everest, they soon after disappeared into the cloud and were never seen again.

Four of the wintering party, Mike Gill, Mike Ward, Wally Romanes and Barry Bishop, reached the 22,000-foot summit. This was a preliminary to a nerve-wracking descent made all the more difficult when a Sherpa broke his leg and had to be carried to safety, much of the way by the summit party climbers Ward and Gill. Ward, a surgeon from London Hospital, splinted the leg with an ice-axe and pieces of a cardboard food box, deadening the pain with morphine before they began the long haul down. After four hours, they were only halfway down their descent and decided to camp overnight, building a makeshift shelter. They were short of food and had no fuel. Surface snow made progress slow next day, forcing another night on the mountain before Sherpas, including a strongman called Karma, arrived with relief and supplies. Karma carried the injured man to the safety of the Silver Hut.

It was a triumph which produced a bureaucratic blizzard. Ed Hillary arrived back from a supply trip to find a letter at the British Embassy at Kathmandu. It was from the Nepalese Officiating Chief of Protocol. Without mincing words, he made it plain that the authorities did not share the enthusiasm about the scaling of Amadablam.

Mr N. M. Singh spelled out the protocol. The party had been given permission to carry out scientific work in the area and to make an attempt on Makalu, but there was no permission given to attempt Amadablam. 'It is regrettable,' the man said, 'that a party led by so experienced a mountaineer as Sir Hillary should have disregarded the rules. As His Majesty's Government believe that for the proper development of international mountaineering in Nepal it is necessary that the regulations are observed strictly, they feel compelled, though reluctantly, to withdraw the permission granted to make the assault on Makalu . . .'

If the wintering party had had problems descending Amadablam, then Hillary's slow climb up the face of Nepalese bureaucracy afterwards was even more difficult. Nine days and a token fine later and the Makalu show was back on the road.

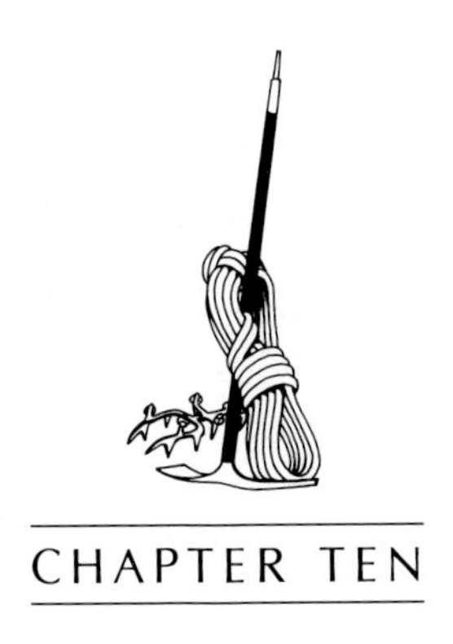

CHAPTER TEN

Disaster on Makalu

Makalu stirs those who view it. That was clear to me that first afternoon in 1952 in Dan Bryant's office when he swung around in his chair and stared at the photo on his wall. It was almost as if his eyes narrowed from the reflection of centuries of snow as he spoke of the peak. He told me of Mallory's description: 'incomparable for its spectacular and rugged grandeur'. If it was second only to Everest in his eyes, then not much separated them. Often during that afternoon, he came back to Makalu. 'Some day,' he said, 'someone will climb it.' Almost regretful that it would not be him. And he had spoken of Shipton back in the area and of the young New Zealander with him.

The young Mike Gill saw it in 1961 from the summit of Amadablam. The leonine outline of Makalu, he called it, and then that moment just before they began their descent:

> Directly ahead loomed the colossus of Everest, no longer squatting behind the Lhotse-Nuptse wall, but for the first time massively dominating the whole fantastic landscape.
>
> To its right, stood Makalu, gracefully proportioned despite its bulk, and in between a vista of rolling brown hills stretched to the horizon where a shaft of light through the clouds played on the snows of a range far inside Tibet.

For some minutes they had stared at the north ridge of Amadablam where the British climbers Harris and Fraser had disappeared little over a year before. Gill says: 'We were appalled at the steepness of the final ice ridge and the ferocious severity of the knife-edge rock falling away below it. Why had they chosen this route, we wondered . . .' Before departing, they took a last glimpse of the peaks around them, especially Makalu.

With Amadablam behind them, the next challenge was Makalu, all 27,790 feet of it. The long routine of their individual and shared research was over. So was what Mike Gill described to me as a regular routine to prevent the melting snow undermining the foundations of the Silver Hut, raising the risk that it might become a giant toboggan sliding off down the slope. There was too the need to wave off what had been known as the Ladies' Expedition, led by Louise Hillary and including June Mulgrew, Irene Ortenburger, Betty Milledge, Lila Bishop and Gita Bannerjee, all party members' wives who had spent some weeks on their own walking expedition in the area.

There was a chilling air of premonition in Peter Mulgrew's description of the takeoff as Louise Hillary and his wife left from the makeshift airstrip at Mingmo:

> The motor was given full power and the plane shook as rocks [under its wheels] were suddenly removed. Like a drunken duck the aircraft ran uncertainly along the rough surface, gathering speed as the wheels struck hidden holes. I felt physically sick as the pilot, at the last minute, pulled back the stick and flung the plane into space over a steep precipice . . . relieved I watched as they flew over Thyangboche monastery . . . and set course for Kathmandu.

Despite the lengthy acclimatisation and the theory it would help — 'allow you to race up Everest or anything else in sight' as Mike Gill described it — the party soon felt the familiar effects of altitude as the first stages of the assault went on. First, it was Sherpas, hard-hit by headaches and vomiting, several of them forced to retreat to lower levels. Then, it was Ed Hillary himself.

Others in the party heard a faint voice in another tent and found Ed Hillary sitting up in his sleeping bag unable to speak.

'A transient aschemic attack' was how Mike Gill described it to me, a shutdown of a blood vessel or where a clot had dropped off, affecting the brain so that he wanted to say words but the wrong words were coming out. In a detailed medical assessment later, Dr Mike Ward described Hillary as having had a transient stroke. He reported three other cases involving young, fit men on Himalayan climbs. Two had died.

Unbeknown to anyone, Ed Hillary had not felt well for several days and typically had not told anyone about it. He later described headaches and back pains and then pains in his head, which worsened. Rushed on to oxygen and sedated, he improved enough to make his own way down,

Sir Edmund Hillary, drawn and ill after his stroke on Makalu, making his way down to lower altitudes.
NEWS MEDIA

helped by Sherpas. He was markedly better at lower altitudes, enough to make the fifteen-day walk back out of the mountains. It was, Mike Gill said, very similar to an attack that Eric Shipton had on Everest.

> If Hillary had carried on perhaps nothing would have happened, but the doctors who were there, Dr Michael Ward and Dr Jim Milledge, a physiologist from the British Medical Research Council, spoke to him very sternly and told him he had to get out of there. And they got him out good and pronto. He was dispatched straight out of there at a height of about 21,000 feet and shot off down the valley to 12,000 feet. He spent the rest of his time walking rather despondently through the foothills making his way to Kathmandu from a very remote part of Nepal.

With Mike Ward now leader, the party began positioning itself for the assault. Used to most of the rather bizarre ways of the sahibs, the Sherpas could hardly believe it when the ergometer bike was dismantled and taken with them to continue the familiar bike-till-you-drop experiments high on the Makalu Col.

The problems of beating the bike were, as it turned out, nothing compared to what lay ahead in the next few days. Over-exertion in what were clearly too exhaustive physiological tests high up the mountain left Mike Ward vulnerable to extreme effects of altitude. He developed a chest infection and began to hallucinate, became extremely weak and sometimes delirious. His fingers, toes and nose were frostbitten. More importantly, his heart was strained and enlarged — it would take three months to revert to normal. He was coughing up blood. When he finally left the mountains he was two stone below normal weight. Little wonder that, in Mike Gill's words, the new leader 'did not make good decisions'.

The assault went ahead in bad weather when, as Mike Gill says, good conditions were really essential, in particular for a climb without oxygen. The result, in Gill's words, was that 'the whole thing came to grief'. As a member of the first assault team, he experienced the danger and the failure at first hand. With him were Wally Romanes of Hastings and the American mathematician/climber Leigh Ortenburger and four Sherpas.

Well, four Sherpas at least to begin with. After what Mike Gill remembers as an endless night in tiny assault tents, the party woke in high winds and clouds of powder snow. Doubting that they could reach the summit but mindful that two other assault parties were queueing behind them to use the camps, the three climbers pressed on towards what would have been Camp Seven at 27,000 feet. That was when one Sherpa reported himself as ill and refused to go higher. He probably knew something and was not prepared to be part of what he felt could be a disaster.

By early afternoon, they were in a blizzard with the risk that they could become storm-bound long-term — if not permanently — at around 27,000 feet. They left the supplies for Camp Seven in a dump and began a perilous descent. Once, Gill and Romanes fell together fifty feet, fortunately into soft snow, after Gill lost his footing.

At last back in their tiny tent, Gill, Romanes and Ortenburger waited fourteen hours before they could move again. Gill's toes and fingers lost all feeling. When they finally set off for Camp Five, he frequently had to stop and rest as lack of oxygen left him panting and weak. Sometimes, it was every hundred yards. Later, sick and disoriented, they listened over the radio link as Peter Mulgrew and Tom Nevison began their ascent.

The shocking misfortune that had been dogging the party stayed with the second assault team. A sudden loss of balance by one of their lead Sherpas whipped him and the five others roped to him down the ice face, falling 600 vertical feet in a slide which carried them 1000 feet. In the seconds, then minutes, that followed, both Mulgrew and Nevison feared that all were dead. When they reached them, they found only two had been injured, one with a damaged leg, the other with head and face cuts from flying crampons.

This was a crucial moment for the Makalu team. Realising that the injured men would have to retreat down the mountain for treatment, Mulgrew and Nevison, an American doctor from the United States School of Aviation Medicine, had to re-assess their situation. They knew that if they turned back, without establishing base at Camp Seven, the team to follow — John Harrison of Christchurch and Ortenburger, who had suffered less badly as a member of the first party — would not make the summit either.

Mulgrew and Nevison decided to press on, with Mulgrew taking most of the load left by the injured Sherpas and with Nevison leading and cutting the way. At just over 27,000 feet and only 800 from the summit, they pitched tents and waved all but one of the Sherpas off to descend. There were no arguments. They were more than happy to go. Annulu remained. The climbers could look across at Everest and trace the route Hillary and Tenzing had taken to its summit, some 2000 feet higher than their camp.

Next morning, they began a literally painful foot by foot ascent towards the summit of Makalu, achieving at best 100 feet an hour, having to pause often to rest, everything a struggle. Slow step by slow step towards the summit. With only 350 feet to go, disaster struck. Peter Mulgrew suddenly plunged forward, his face contorted with pain, bleeding as he breathed, each breath an agony. He had difficulty speaking. When he sought to stand, another terrible stab of pain felled him to his knees again. The pain was constant. His breathing was laboured and a desperate struggle.

He knew he was literally in mortal danger. At the height they were, in those conditions, help was perhaps days away and it was clear that he could climb no higher. Typically, in a gesture Captain Oates would have understood, he suggested his companions go on to the summit and come back for him.

Nevison and Annulu too knew the danger and would not leave him. Mulgrew had suffered a major clot in his lung, a massive version of the much smaller brain clot which had caused problems days earlier for Ed Hillary. Where Mike Gill later described himself to me as having reached the limits of his endurance, he said of Mulgrew at that stage that he was 'down to the last cylinder of his capacity'.

The bid to climb Makalu was by then a write-off. It had instead become a desperate attempt to get off the mountain without loss of life. In fact, all three members of that party on the high slopes were in trouble, Mulgrew worst of all, but the others with handicaps too. Nevison diagnosed himself as having an attack of pulmonary oedema. His sputum was tinged with blood. Unknown to either of his companions, a quiet but stoical Annulu was in severe pain with a cracked rib from an earlier fall, an injury he bore without a murmur.

Slowly, often only a few yards at a time before he crumpled to the snow again, his body wracked by new pain, Peter Mulgrew worked his way back with his companions to their last tent. There, Mulgrew lay awake, unable to sleep with the agony of his condition and wondering whether he would survive. Next day, having left their tent behind, they attempted to make their way down to Camp Six. The weather was bad, with high winds and little visibility.

After Mulgrew collapsed again, it was obvious that they would never reach the lower camp without help. While Annulu, still hiding his injury, went on for assistance, an exhausted Tom Nevison dug out an ice cave as shelter for the night. For the first time, as well as the severe pain in his chest and side, Peter Mulgrew noticed that he had lost all feeling in his feet. As their condition worsened and the weather continued its clamp around them, even men as determined as these two began to lose hope. With their minds drifting, they wondered whether the rest of the party had perhaps already given them up for dead and begun to retreat down the mountain.

The arrival of two Sherpas, Pember and Pasang Tenzing, dispelled that fear and gave them a reserve of oxygen. Nevison himself was by now so ill that he realised he could not remain without risking collapsing. So, when a gallant Leigh Ortenburger arrived with more oxygen, Nevison regretfully but sensibly made his way down, leaving his fellow American to cope.

Although somewhat better with the oxygen boost, Mulgrew was still hallucinating, believing himself at one point home in New Zealand at the beach with his family, at other times in terror over some perceived but non-existent menace which outweighed even the terrible peril he was in. Frequently barely conscious and at best able to make perhaps half a dozen steps, Mulgrew set off with Ortenburger for Camp Six, and then with the aid of the Sherpas to Camp Five, collapsing often, drifting into delirium and finally reaching a stage where he could not even stagger. Sherpas carried him from above 25,000 feet to the lower camps.

Peter Mulgrew, his frost-bitten feet and hands heavily bandaged, with his wife June and Sir Edmund Hillary in an ambulance in Auckland on their way to the Devonport naval hospital after arrival at Whenuapai.
THE PRESS

There, his New Zealand comrades saw him for the first time. They were appalled. Mike Gill remembers: 'He had changed beyond recognition. He was gaunt and ashen grey with sunken eyes and frostbite showing as black patches on his face and hands. His feet deceptively looked a pale lilac which later ominously turned black.' Carried over the final stages in a roughly made sledge stretcher, Mulgrew knew nothing of the tremendous struggle to save him. He had virtually lost the use of one lung. Many times, the party stopped and his pulse was checked to assure his rescuers that he was still alive. He was — only just.

Fellow New Zealanders Wally Romanes and John Harrison — who was slogging on despite frostbitten feet — were part of the team inching

Prime Minister Keith Holyoake presents Peter Mulgrew with a cherrywood walking stick in October 1961. Within two months, Mulgrew was using the stick to walk.
NEWS MEDIA

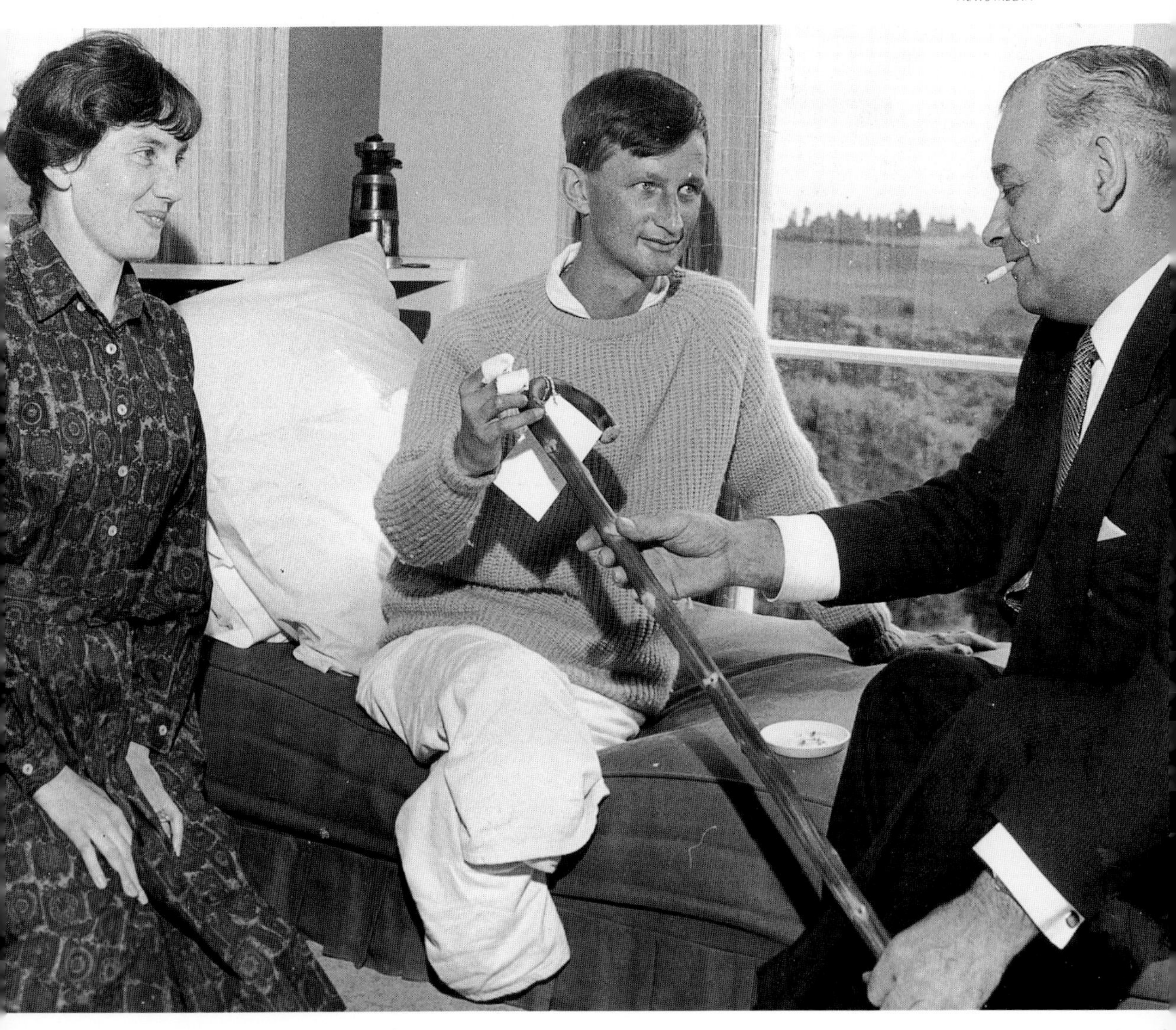

Peter Mulgrew, now with his artificial legs fitted, with his wife June and daughters Robyn and Susan, rebuilding their lives after the Makalu disaster.
ROBYN MULGREW

the sledge which Harrison had lashed together out of pack frames, tent poles and rope down the slopes until, below Camp One, an American helicopter lifted Mulgrew off and headed for Kathmandu.

Fortunately, with them on the descent was another of the party's six doctors, John West from the Postgraduate School of London and the British Medical Research Council. They all had a role. Clearly though, in a mission with many heroes, Leigh Ortenburger had been magnificent. Said Mike Gill: 'I'm not sure Peter would have made it but for Leigh.'

Bad as those final days on Makalu had been, they could have been much worse. The loss of Ed Hillary, both for his leadership and his sheer physical presence, had been crucial. Then, the almost total loss of deputy leader Michael Ward had dealt a second significant blow. With Hillary and his escort of Barry Bishop and Dr Jim Milledge not available, other climbers had been forced to carry unexpectedly large loads at altitude as the assault team positioned themselves. The accident involving the six Sherpas and the injuries there made that worse.

What was also clear was that the theory tested in the long winter acclimatisation was not valid. The men who had wintered over showed no benefits. In fact, the lack of sustained hard exercise at 19,000 feet around the Silver Hut had left them ill prepared for the difficult conditions they faced on the upper slopes of the mountain later.

Significantly, John Harrison and Leigh Ortenburger, who were comparatively late arrivals and not in the wintering party, were strongest, with Ortenburger's physical condition and determination a crucial factor in the battle to save Mulgrew. Medical assessment later showed that the wintering party was tired and jaded, that 19,000 feet was much too high for maximum acclimatisation.

Dr Michael Ward, who had himself suffered badly as a result, was

Only four months after amputation, the indomitable Peter Mulgrew is upright again on temporary artificial legs.
NEWS MEDIA

emphatic later that only the presence of oxygen prevented the probability of three deaths in those grim days — certainly Mulgrew's rescue bordered on the miraculous, and Nevison and Ward were both in potentially life-threatening situations. As an indication of his state, Mike Ward took five and a half hours of grim struggle to move from Camp Five to Camp Four — a journey a far-from-well Jim Nevison had covered in a quarter of that time. And Ward only just made it.

Ed Hillary quickly recovered from his setback, Ward and Nevison took longer but also worked their way back to full health again.

For Peter Mulgrew, rescue from Makalu and the end of his battle to survive there was the beginning of months of struggle. Reunited with his wife June, who flew back to Nepal to be at his bedside, he faced months of pain there and on his return to New Zealand until, inevitably, his lower legs were amputated. He had, in the meantime, battled through a pethidine addiction — he had two-hourly injections of the pain-killer for months to make his life a little easier — and then had to struggle to master artificial limbs. He applied to this new struggle all the greatness of spirit and raw determination which had sustained him on Makalu. Those who knew him well were not surprised either at this will or where it took him.

Old comrade Ed Hillary, who had seen him through the rigours of the Nepalese hospital spell and the journey home, called one day with a copy of Douglas Bader's life story. The message was obvious. If Bader could recover from the loss of his legs in a pre-war air training crash and go on to not only walk and fly again, gaining fighter ace ranking in wartime and becoming a legend too, then Peter Mulgrew could do the same.

Mulgrew had the will long before he read the book. From first clumsy steps and endless progression around the family furniture, he learnt to walk again — and more. A little over two years later, he was back in the Himalayas with Ed Hillary's schoolhouse-building expedition.

He also built himself a new career in commerce, and became a skilled yachtsman, sailing for New Zealand in One Ton Cup international competition. By any standards and in so many ways, Peter Mulgrew was a most remarkable man. Even set against the historic qualities of men with whom he shared his life, his courage and never-failing determination were outstanding. So was the loyalty he showed and shared.

His link with Ed Hillary, forged in times of triumph and disaster, was to outlive double tragedy.

CHAPTER ELEVEN

Journey to the Gods

Ancient wisdom sometimes comes from unexpected sources. One day late in 1968, Ed Hillary was given such advice. He and his party were on the banks of the Arun, one of the big rivers in India's Ganges waterway complex. As at so many stages of his life, Ed Hillary was testing out a new, adventurous theory.

The previous year, in the spring of 1967, he had achieved another of his objectives, putting New Zealand climbers on the summit of 11,000-foot Mt Herschel, an unclimbed peak in Antarctica, as part of a five-week expedition there. Familiar names travelled south with him — Norm Hardie, an old Himalayan teammate and conqueror of famed Kanchenjunga, Murray Ellis, one of the New Zealanders who went to the Pole, Mike Gill, by now a Himalayan veteran, and three other climbers, Bruce Jenkinson, Dr Peter Strang and Mike White. The latter four all climbed Herschel.

As with all his close companions in adventure, Ed had a special affinity with Mike Gill. He said of him later: 'If an Antarctic blizzard is trying to demolish your tent, or you're camped on an uncomfortable ledge at 20,000 feet and don't know whether to go up or down, then Mike is a magnificent companion.' It was a feeling that was reciprocated. It was not surprising then that Mike Gill got another of those famous throw-away phone calls from Ed Hillary to join him on his next adventure.

Hillary had been fascinated for some years by the huge rivers that flowed from the Himalayas, some direct from Everest itself. It was more than the spur of challenge. He believed they could be a valuable link with the high country of Nepal. Having ridden New Zealand designed jet boats

Sir Edmund Hillary in one of the Indian saris he wore almost like a uniform on the Ganges trip.
NZ HERALD

with Jon Hamilton, son of their inventor, he was determined to put the two components together.

That was what took him to the banks of the Arun, having set up a base near where the three mighty rivers, the Sun Kosi, the Arun and the Tamur, met. The prime aim was to attempt to reach Kathmandu by water up the Sun Kosi. Two Hamilton boats, aptly named *Kiwi* and *Sherpa*, were to make the trip. A first attempt at the Arun produced that ancient local wisdom and an interesting outcome. Mike Gill remembers the wisdom, the source and the sequel very well.

'We met this little Nepali guy. He was just walking along as they do, very brown and wearing very little, slim little men who have been padding around the hills all their lives. We told him we were going to this big river. And it *was* big and quite fast, bigger than anything in New Zealand, bigger than the Clutha by quite a long way.

'He immediately looked pretty alarmed and made it clear that we

Jon Hamilton, Edmund Hillary and the Sherpa Mingma during preparations for the pioneer journey up the Sun Kosi and Arun Rivers, the preliminary to the attempt on the Ganges.
THE PRESS

should not do that. We explained through our Sherpas that we had good boats and we were skilled and wouldn't have any problems. He wasn't convinced.

'That afternoon, we headed off. It got bloody difficult to say the least. I had taken the precaution to go with Ed and Jon Hamilton, which was very wise. The other boat was being driven by Jim Wilson, who had comparatively little experience and he was obviously having a pretty hard time.

'Finally after we had been going a couple of hours, with some big rapids and very large waves, I looked around to see the stern of the second boat sticking straight up in the air blowing up a huge stream of water like a giant whale sounding. We turned and raced back. Jim Wilson, Max Pearl and the Sherpa Ang Passang — he was totally terrified and has never been near a boat or the water since — were floating down the river.

'After we rescued them, some of us were walking down the river when we met this little Nepali guy again, walking up. Under his arm, he had a seat out of the jet boat, and he was wearing a very, very large smile.'

It isn't clear whether there is an equivalent of 'I told you so' in Nepalese, but if there is, that's what he would have said. Ed Hillary's reaction to the loss of the boat? Merely a shrug, a smile and some Auckland Grammar French — 'c'est la vie!'

Shaken, but still basically undeterred, the party moved on to the Sun Kosi with the remaining *Kiwi* to attempt that journey all the way to Kathmandu. It was a trip that blended grandeur with a nightmarish component of fiendish rapids and rocks, forcing them sometimes to strip the boat of all gear to allow Jon Hamilton to crash through the worst obstacles alone. It was a 250-mile, two-week epic.

Around them were cliffs and mighty forests, valleys that led away from the river, villages that had never seen a white man before, much less a jet boat, days of excitement until that moment when they eased into Kathmandu itself. They had proved Ed's point — and unwittingly set another challenge. Ed was hooked on the thrill of it all. He wrote afterwards: 'None of my adventures have left me with more exciting and enjoyable memories, and I have a feeling that the rivers of Nepal will see our jet boats again.'

Nine years later, Ed Hillary, the jet boats and a new ambition were back. This time, it was an even greater challenge — a journey from the mouth of the Ganges to the Himalayas where the sacred river first comes to life. When Hillary called it Earth to Sky, he summed up both the geography of it and a very special spiritual aspect which was to affect so many of the party deeply and long term.

Even now, Ed Hillary, so much the matter-of-fact person he has always been, talks with an obvious reverence of the river as 'Mother

Ganga', as a fervent Hindu might. I can still remember the special tone to his voice when we first talked of this journey. To him it was clearly much more than a typical adventure, there was a sense of pilgrimage and discovery which remains still.

It was more than simply a journey to the beginnings of a sacred river, there was, too, a return to his own past. The 1500-mile voyage would seek to end at Badrinath, where a much younger Ed Hillary had set out on his first Himalayan expedition back in 1951, and where he later received the telegram from Eric Shipton inviting him to join the 1952 Everest expedition that was to alter his life.

Those who travelled with him also brought with them parts of that new life which Everest had shaped for him. Inevitably, there was Mike Gill as deputy leader, now forty and, despite the constant, almost annual, lures of Hillary expeditions, an experienced doctor as well as a fine climber. Jim Wilson had first gone to the Himalayas with Hillary in 1963, newly graduated as an MA in philosophy. Afterward he and his wife Ann settled in Benares. There, he earned a doctorate based on his study of Hindu beliefs and philosophy. He loved and knew India, its history and its mystery. He was, as well, a splendid climber and an excellent canoeist.

Twenty-two-year-old Peter Hillary was a fine climber, son of a great climber, grandson of another, already with an impressive list of climbs to his credit and with others to come. Murray Jones (32), another Himalayan veteran, had, with Graeme Dingle, made the first New Zealand ascent of the north face of the Eiger and numerous other notable firsts. Dr Max Pearl (52), had long been a keen worker for Hillary's Himalayan Trust, dedicated to educational and health work for the Sherpas.

Then, there was Graeme Dingle (32), one of the great characters of New Zealand climbing and exploration, who had already climbed with Murray Jones, had made the first traverse of New Zealand's Southern Alps and been with a New Zealand expedition to the north face of Jannu in 1975. He would later team with Peter Hillary in an historic Himalayan traverse and, in 1993, would head back to the Arctic to complete a 16,000-mile journey around the Arctic Circle.

Dingle, then 'a scrawny young climber', as he describes himself, first met Ed Hillary when the group for the Herschel climb were in training at Waiouru. The meeting was memorable for several things. Graeme Dingle was nursing a 'smashed' left arm, injured in a fall, and he had with him a very famous Italian climber called Carlo Mauri, who had among other feats climbed one of the toughest Himalayan peaks, Gasherbrum 4, with another noted Italian, Walter Bonatti.

Ed Hillary was there, Dingle remembers, with Peter Mulgrew and Mike Gill, 'who were sorts of demi-gods', all standing on the steps of a barrack building when Carlo Mauri began pushing a very shy Graeme

Dingle forward with a mission. 'Go and ask Sir Hillary if I can go with him to the Antarctic.' A bashful Dingle made the introduction. It was quite clear that Hillary had never heard of Mauri, but recognised the name Bonatti. The young Dingle remembered Hillary at that first meeting as rather off-hand — which he can be. The meeting ended without an invitation to Mauri, who later got to the ice, but not through Hillary.

Later, the Dingle-Jones feat of climbing all the major north faces in Europe drew an invitation for them to join one of Hillary's parties in 1970 to climb a Himalayan peak called Karyolung. This time, a more relaxed Hillary impressed the new man. Finally, Karyolung was never attempted by the party. Permission was withheld — 'it was kept for a Nepalese party to climb later' — but Dingle and Gill made a reconnaissance of the slopes on the south face to nearly 20,000 feet.

Instead, in a pattern which had become typical of Hillary expedi-

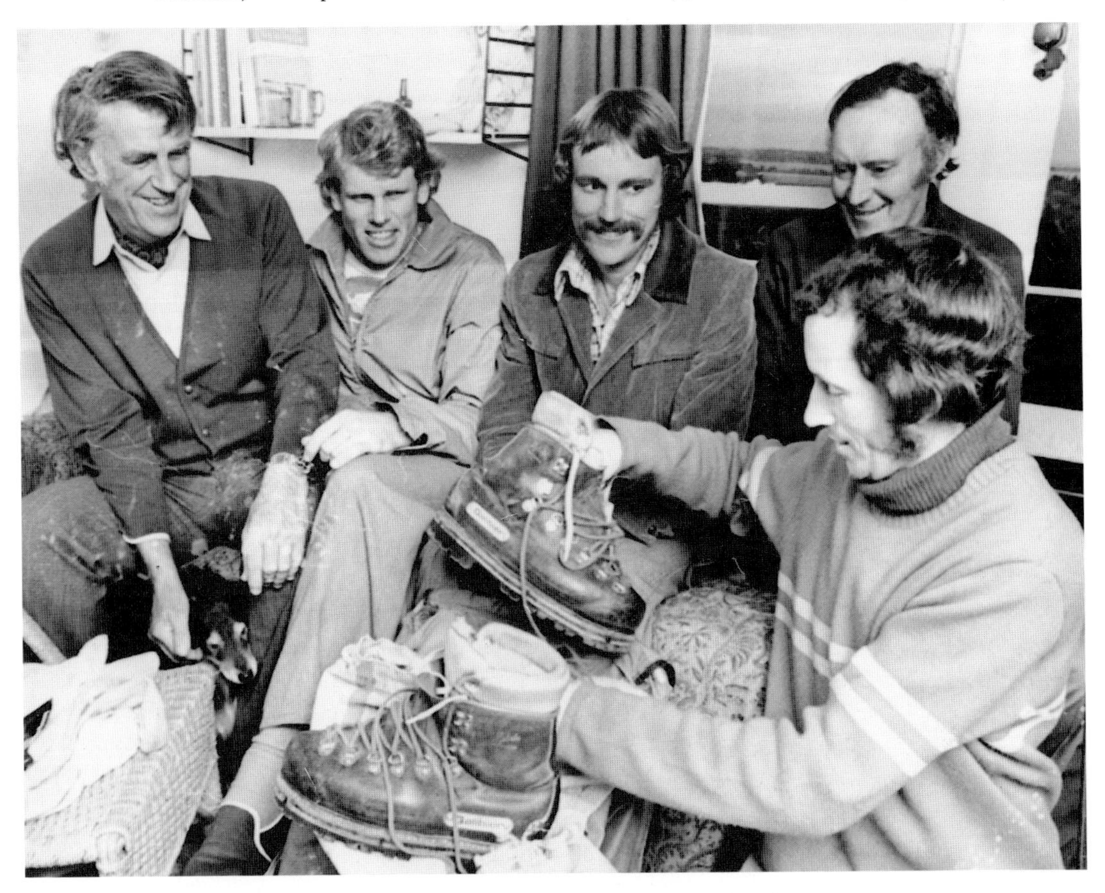

Edmund and Peter Hillary with the jet-boat experts, Michael and Jon Hamilton, and Graeme Dingle (with boots) just before they left on the Ganges adventure.
EVENING POST

tions, the party spent its time doing what Dingle describes as 'a huge amount of work — it's hard to believe just how much we did — building two bridges, one complete school and modified two others substantially, major work to the Khunde hospital, built a huge extention on to the Thami terrace fifty or sixty feet high in solid stone, put in a water supply to Thyangboche monastery, investigated a supply for Khunde . . . it just went on and on.'

Graeme Dingle was packing to leave for the Arctic as we talked, a process interrupted by a phone call from Ed Hillary to wish him luck. That call brought back memories, in particular of what was probably Ed Hillary's last major climb of a big mountain, a grand traverse of Mt Cook, New Zealand's highest peak, in January 1971.

It was an historic party, three generations combining: Harry Ayres, who had played such a part in the shaping of Hillary the climber, Ed Hillary, Mike Gill, Jim Wilson, Graeme Dingle and Lyn Crawford. More than twenty years later, Graeme Dingle took time off from last preparations for his latest adventure to recall it.

'. . . an exceptional trip . . . a lovely outing, spoiled only by us finding the body of a lost climber on the way down. That climb was very special.

'Even then, Ed was still clearly very strong, an exceptionally strong person, but not a technical sort of climber. He probably couldn't have done some of the hanging from the fingers we do on some of the very technical sort of stuff.

'Remember that, the day before the summit on Everest, he had carried nearly eighty pounds to around 27,000 feet. Unbelievable strength and he would throw in a couple of cans of peaches [his favourite climbing food]. It doesn't sound much at sea level but up there people cut off the handles of their toothbrushes and stuff like that. He had the sort of strength that, to a certain extent, New Zealand climbers have lost.

'As a person, he had this ability to slowly absorb information and then use it to make the right decisions. I saw that plenty of times.

'He could be bloody blunt and even rude at times. I remember once when I was playing football with a lot of the Sherpas on a flat below Namche Bazaar, I punted the ball about a hundred feet in the air and to my absolute horror the thing came down on the top of Ed and Louise's tent. He came rampaging out and roared at me that I had already wrecked one tent, now what was I doing? He was really angry but a few minutes later he walked past and gave me a bit of a wink.

'The other tent incident had involved one of Ed's beloved Sears Roebuck tents being used by a couple of our friends. I wandered over to it to push the flexible fibreglass pole over on top of the people sleeping inside but unfortunately the pole broke and I fell on top of it. There was nothing Ed loved more than his tents. "That's my tent you've just wrecked, Dingle."

'I remember Mike Gill saying to me that it was certainly the last Hillary trip I would go on. I think I was a frustrating bugger for Ed, always breaking things or being a bloody nuisance.'

Mike Gill tells his own story of the Sears Roebuck gear, which Ed Hillary had a contract to field test. There was, Gill says, a rather Father Christmas atmosphere about the arrival of the test gear and Hillary's distribution of it. 'He was very protective about it. There'd be thermal underwear, jackets and boots. I remember once I got this very good set of thermal underwear and there was this elderly Tibetan there eyeing it, and he had a good carpet.

'Well, he and I talked about this for about an hour and finally I traded the underwear for the carpet. I thought I'd got a good deal. He thought he had got one. But Ed was absolutely ropable. I think he had visions of having to go back and tell Sears Roebuck that one member of the party had discovered in one hour that a Tibetan carpet was more valuable than their underwear.'

Obviously, Ed Hillary bore no grudges, because in 1977 they were all there again for Earth to Sky — Hillary, with a new batch of gear from Sears Roebuck, Mike Gill and Graeme Dingle among them.

Graeme Dingle has no hesitation in defining it. With all he had done, has done since and plans to do, that was the best-ever expedition . . . 'the total adventure, a complete experience, the cultural and spiritual, all together. The intensity of it built up as the trip went on. For me, it began an incredible love affair with India that I had trouble breaking away from. I simply had to go back to India every year for the next decade.' Graeme Dingle, exuberant and often noisy and unpredictable, was not alone in this feeling of having shared a great spiritual experience.

With three jet boats and with experts Jon Hamilton and his son Michael in the party, the plan seemed a simple technical feat and adventure — to navigate up the Ganges. It was that, but much more. The river has immense religious significance to Hindus, and this sense of a national, shared and personal pilgrimage quickly gripped the party.

This added spiritual dimension was all around them virtually from the first day on the river, when, at the temple of Ganga Segar, an old priest, chanting and with bells ringing, and with the party in procession, blessed their jet boats in a marriage of ancient ritual and modern technology. He broke coconuts over their bows, placed a dab of colour on the forehead of each of the party and on the bows of the jets, adding garlands of flowers. Then he filled a copper container with Ganga *pani*, the water of the sacred river, to travel with the adventurers until the end of their journey where the great torrent began its life.

The expedition boats, garlanded and blessed, ready for the long journey to the source of the Ganges.
THE PRESS

What had begun as an adventure had in that hour became a very special pilgrimage. This remained and deepened in the weeks that followed as the boats moved up the river, to be met and acclaimed by millions of Indians on the river banks as they did — in their hundreds of thousands in the huge cities like Calcutta, with awe in tiny villages, with reverence in the various holy places on their route.

None of the party ever forgot those days on the river, Ed Hillary least of all. Any later discussion with him on belief or religion inevitably involved those times of unexpected emotion on Earth to Sky. When we talked about it one day during an interview years later, he fell silent for several minutes, saying nothing as he remembered it all. Then, he began slowly, and with still obvious respect, to retell the account of their typical arrival at Varanasi, Benares, the chanting, the bells and the flaming lamps, the flowers. The impact of it is still with him.

Jim Wilson, with his deep understanding of Hinduism, saw that occasion and others like it for what they were — Edmund Hillary, Man of Everest, being welcomed as a virtually divine being, making a pilgrimage which the ordinary people would have felt was the pinnacle of their religious life. By being in his presence, even touching him, they gained a share in this. This was what drew the masses of the huge, poverty-ridden cities, the farmers from their fields, the fishermen, the women with children on their hips, to the banks of Mother Ganga and her many holy places as these men from a distant place jetted through.

On they travelled, through the vast spaces of the river, through gorges and rapids, until finally, just over a month after those first prayers at the river mouth, they found their way blocked by an impassable vertical, ten-foot waterfall at Nandaprayag. They were 1500 miles from the sea.

Jon Hamilton guns one of the jet boats along a channel of the Ganges delta.
THE PRESS

Modern jet-boat meets ancient dhow on the Ganges.
THE PRESS

Edmund Hillary is an anxious passenger as Jim Wilson steers through rapids on the Ganges.
THE PRESS

Kiwi ingenuity prevails. With an aluminium strut damaged, Edmund Hillary fits a bamboo replacement.
THE PRESS

Now on foot, they trekked to the sacred lake of Hemkund up the 1,064-step Golden Staircase and on to Badrinath, with all its memories of that long-ago call to Everest for a young Ed Hillary, and where he and George Lowe had explored in those first Himalayan days in 1951.

Times were different now, and Ed Hillary was quick to notice it. Ahead lay another potential disaster. What exactly happened in the build-up depends on who you are talking to. Graeme Dingle has no doubts. The party went to altitude too quickly.

'Ed was a bit bullheaded about the speed of our approach to the final climbs. Obviously at fifty-eight not a young man any more. I remember as we were walking up the road I said to him: "Hey, this programme we've got is a bit fast isn't it?" I know that Mike Gill, at least, had already spoken to him about it as well. Ed was pretty off-hand. He said we just didn't have any more time.

'So, I remember on the first day we carried pretty heavy loads from 12,000 to about 15,000 feet to set up a base camp. Next day, we carried loads up to about 18,000 and more loads next day and stayed in that camp. Ed wasn't looking that good from the time we got there, and slowly over the next few days got more sick.'

Ed Hillary acknowledges that there was too much pace on, but, interestingly, lays the responsibility at the boots of his younger climbing mates. 'I had planned to move very slowly above base camp to give time for acclimatisation, but foolishly I allowed myself to be over-ruled by the more vigorous members of the party.' (Since Dingle, who fits that description, says he was warning against such speed, and apparently Mike Gill was too, it's questionable who this strong-willed influence was.)

Ed Hillary: 'We were driving desperately and unnecessarily on, although many of us were clearly showing the effects of altitude. Let's get it all over and done with, seemed our aim.'

The intention was to climb two adjacent peaks. The first target, Narayan Parbat, had been investigated to 17,000 feet by Peter Hillary and Murray Jones. It was, they said, much too difficult. Instead, the party planned to climb Nar Parbat and nearby Akash Parbat, just over 19,000 feet.

Even Peter Hillary had felt the effects of altitude on the reconnaissance. Familiar problems had begun for his father as well. He was having trouble coping with the effort involved in the climbing and the carrying. He wasn't eating and his sleep was wrecked by nightmares. His head and back ached. Mike Gill remembers Hillary complaining of bad back pain and giving him a pain killer. Next morning, Hillary was unconscious and could not be brought around.

Mike Gill talked later of a 'panicky sense of urgency'. Altitude sickness can be a killer if it is not responded to with oxygen or a drop in height. Several people a year, on average, die in the Himalayas from this cause. They don't make headlines. They are not Ed Hillary. Graeme Dingle remembers: 'We figured that he was in real trouble and might not have long to live. We collapsed the tent around him in his sleeping bag.'

The party tied ropes to this strange makeshift sledge effect of tent and mattresses and began a hard haul down the mountain. Dingle describes them descending about 3000 feet that way in an hour. In the meantime, Murray Jones had left as a runner down the slopes to alert a helicopter.

'He shouted down this decrepit old Indian telephone to the ambassador in New Delhi that Ed Hillary was dying and we needed a copter urgently,' Dingle says.

'After we got off the snow at the bottom of the glacial area, we tried to carry him. I remember his legs dragging on the ground. I weighed about ten stone and he must have weighed about seventeen and a half stone and I can remember saying in his ear: "You're a heavy bastard!" and all the time he was saying "twelve stone five". He was only semi-conscious.'

Down from the altitude, with oxygen available — plus a tin of those favourite peaches which Peter rushed up to his father — the worst of the crisis was over. After a brief spell in a military hospital, where tests showed a strained heart and that familiar pulmonary oedema, Hillary returned to Badrinath, having in the meantime politely waved away the Indian president's personal jet sent to ferry him south if needed.

He was there in time to welcome back successful assault parties. Murray Jones, Peter Hillary and Graeme Dingle had climbed Nar Parbat; Peter Hillary, Jim Wilson, Dingle and the Sherpa Waka had climbed Akash Parbat. The water from the mouth of Mother Ganga, carried so carefully on that long spiritual pilgrimage, had been sprinkled on the summit snow.

The Journey to the Gods was over.

CHAPTER TWELVE

Paradise Lost

They were a few words in thousands Ed Hillary said and wrote about the Earth to Sky journey, but they mirrored so much about the years before. He was describing the reception in Varanasi, the fervour, the colour, and how later he had gone back to the expedition camp: 'I felt stimulated and revived; for a time at least the sadness of the previous two and a half years had vanished from my mind. Life was worth living again . . .'

This was very much the way he spoke to me years later when I interviewed him for a profile to be included in a gallery of New Zealand Living Treasures in *North & South* magazine. We had been talking about the years of joy in the life of Hillary — and a decade of agonising sorrow that followed. Only then was he able to talk about that 'great shadow', as Mike Gill described it, which hung over him for so long and which had taken him to the depths of despair.

It was somehow so right that they should have always remembered a few moments on a mountain. The young pre-Everest Ed Hillary inevitably was befriended by Jim Rose, for years an enthusiastic climber and by the 1950s president of the New Zealand Alpine Club. They lived in the same Auckland suburb of Remuera and young Ed was often in the Rose home in the years before Everest. He couldn't help but notice Louise, Jim's very lively, always friendly and musically very talented daughter, eleven years younger than him, constantly on the move as late teenage girls so often are.

Notice, it seems, was all he did. He was, as he put it in his own words, 'never very much at home with the opposite sex', although he had been engaged at one time. He and his brother Rex had become engaged, both briefly, to two friends from the Auckland suburb of Sandringham early in the 1940s, a short-lived romance for Ed and a girl the family remembers

Sir Edmund Hillary on Auckland's Mt Hobson — an aura of greatness never left him.
NORTH & SOUTH

Dressing for the part, Edmund and Louise Hillary simulate a Kathmandu marketplace for one of their fundraising events.
NZ WOMAN'S WEEKLY

simply as 'Shirley'. The experience bore out his later assessment that he was not as a young man very relaxed with women.

But he was certainly very much at home on a mountain, and it was on Mt Ruapehu that he first realised that he had become more than simply 'aware' of Louise Rose. Understandably, with her family background, she loved the mountains, but was not an experienced or great climber. That was why family friend Ed Hillary was enlisted to take Louise and some of her friends on a climbing training session up the mountain. Up quickly became down when she slipped and slid directly at her instructor, crampons flashing. When he as quickly jumped out of her path, she sailed past to end smiling and undamaged, looking up at him from the snow when he went to her aid.

That smile was the beginning of a great love. Ed Hillary realised when Louise sailed off to Sydney sometime later, to continue her study on the viola at Sydney Conservatorium, that something had gone out of his life. When he left for the Himalayas later, he carefully allowed himself several days in Sydney to be with her. The significant test of what had happened to him was that Ed Hillary felt real regret at their parting — even though he was going off to what he had until then believed his first love, the mountains of Nepal.

Returning after Everest, he stopped in Sydney — this time to propose, and she accepted. When they married in Auckland a few weeks later, the Hillary-Lowe lecture tour became their honeymoon.

The usually rather taciturn Hillary described their marriage as the happiest twenty-two years of his life to that point. No one who knew them doubted it.

Louise Hillary was so different, younger, bright yet intense, outgoing, musical. He was still the strong, energetic man of the outdoors, good humoured, equally intense about his ambitions, a Joan Baez fan where she played the classics, a Max Brand cowboy reader where her reading took her deeper and further. Yet, it was a perfect match.

His final accolade of her as wife: that she never once said no, or even expressed reservations, when his adventuring spirit took him away yet again — as it did so often during their marriage. Better than that, she was prepared to go with him, children and all — on camping expeditions around New Zealand or across the United States, to Alaska, in the Australian outback, and, of course, to Nepal (inevitably testing Sears Roebuck products as they went).

Following pages: Edmund and Louise Hillary were married on 3 September 1953 — Louise Hillary's twenty-third birthday. The bride and groom left the chapel through an arch of ice-axes.
CHRISTCHURCH STAR

The arrival of Peter, eighteen months later Sarah, and then Belinda, four years younger than Peter, produced a family whose whole way of life revolved around adventuring. Peter recalls as a twelve-year-old climbing with Louise and Ed to the summit of an 18,000-foot Himalayan peak, Kala Pittar, looking across to the majestic bulk of Everest. He realised, for the first time, the literal height of achievement his father had attained — and for the first time experienced the effect of altitude that is such a factor in success or failure. Later, as a world-ranked mountaineer, he knew both.

Louise Hillary, still not a great climber, became as deeply involved as Ed in his aid projects. These began in earnest within a few weeks of Peter Mulgrew's narrow escape on Makalu, when Ed Hillary and several of that expedition remained on in Nepal to assemble an aluminium schoolhouse at Khumjung.

It was very clearly the beginning of a life-long programme to thank the Sherpa people for all they had done, most recently then in the rescue of Mulgrew. Two years later, in 1963, Hillary was back and over the years returned again and again, with water supplies for Khunde village, a schoolhouse for Pangboche, hospitals, clinics, power supplies, air strips. The list of good and essential works seems endless.

And Louise was always a tireless worker, raising money with programmes of fetes and galas in Auckland, leading, convincing and doing. She learnt Nepali. The bookshelves in their Remuera home were packed with references, surrounded by all the mementoes of their second home far away — silk panels, Nepalese hangings and rugs, copper and brass ornaments and containers, intricately carved tables and cabinets.

She taught herself photography — 'so often we need a photographer on Edmund's trips'. She helped handle the hundreds of letters and requests that descended on them regularly. Like the twenty-two-year-old who had packed up her viola one day in Sydney to return and marry, she continued her music, playing in occasional orchestras and maintaining her place in a string quartet. Ed's world was always all around her — the past on Everest, the Pole, the challenges to come.

When Tenzing's daughter Nima visited New Zealand in 1964, a small and delicate figure with waist-length hair, she stayed at the Hillary home. She, like the Hillary children, had been to Everest in her early teens, climbing to 20,000 feet with her father and sister. Later, she was one of a party on Cho Oyu struck by an avalanche that killed four of the expedition, two of them women.

Who else to read the many messages of congratulations at the Hillary wedding but George Lowe?
NEWS MEDIA

The Hillary family welcomes home new baby Belinda in 1959.
NEWS MEDIA

SUPPLIES
BBC
PELMETS
NORCRAFT DOLLS HOUSE

Above: Packing for a long family stay in the United States while Sir Edmund was engaged on a lecture tour in 1961.
NEWS MEDIA

Left above: After opening a handcrafts and hobbies exhibition in London, Louise plays a violin made by George Robey, the famous British comedian.
THE PRESS

Left below: Louise and Edmund with the 1959 Giant of Adventure award, presented at a ceremony in London's Savoy Hilton Hotel.
NEWS MEDIA

When her father Tenzing came to Auckland in 1971, he too was a guest of the Hillary home. He told them delightedly how his son Norbu had celebrated his tenth birthday at 15,000 feet in Sikkim. He said later how the Hillary house seemed more Tibetan than European and talked endlessly of Louise and her kindness. Hillary, a man revered in Nepal, had a wife who was deeply loved there too.

Even a leg broken on a walking expedition in Nepal did not deter her. She joked about the injury giving her a form of membership of the mountaineers' club and she was quickly back on her fund-raising in New Zealand and her health and education awareness drives in Nepal.

Mike Gill, now like Ed past his climbing days, is one of the many doctors who has spent time in the Himalayas in his medical role and is a member of the medical sub-committee of Hillary's Himalayan Trust which arranges the movement of doctors to continue the work there. Another was Dr Kaye Ibbotson, Professor of Endocrinology at Auckland Medical School. He went on a special mission to counter the thyroid problem caused by an iodine deficiency in the soil which is always a problem in mountains. The symptoms are obvious in greatly enlarged thyroid glands. Mike Gill talks of the worst times, now past, when one in ten Sherpa children were born cretins because of lack of iodine. A campaign of iodine injections which Professor Ibbotson set up did all he had hoped for. Even the first round of injections produced amazing results.

It involved a large syringe of rather gluey oil — that's Mike Gill's description — injected into a muscle, producing a reserve of iodine slowly released over the next few years, with the follow-up every three years. Very few cretin children have been born since, and the couple he has noted were born from people who had somehow missed the injection.

There was also a theory that the whole process, once established, was self-generating. Sherpa lavatories are long drops with leaves at the bottom of the shaft. They form big compost heaps that the Sherpas drag out at the beginning of the potato season to fertilise the crop. The iodine content simply goes into circulation again through them when the crops are eaten.

Ed Hillary's right-and-left-hand-man Mingma Tsering had good reason to be grateful. His wife Angduli was a classic example of the problems of iodine deficiency. Her first child was born dead. Her second was badly intellectually retarded. Mike Gill's diagnosis: 'He was retarded to a degree I've never seen anywhere. An absolute cabbage who was propped up in a corner, couldn't move, had to be fed, couldn't do any-

Home at last to the children's pet bantams, December 1962. But there is an obvious problem over who should hold Suzie and Mary, the bantams. Sarah Hillary is on the right, Belinda on the left with Peter making last-minute adjustments.
NEWS MEDIA

The Hillary family with lamas at the Chewon monastery in 1967.
THE PRESS

thing and died at about twenty. Her next child was born deaf. After that the iodine came in and her remaining two children were all right from then on.'

Mingma never learnt to read or write but he had the tremendous ability of Sherpas to retain facts through an amazing memory. He could keep track of 500 porters, their names, what they were to be paid and when. He could pass on the most detailed orders for timber and equipment. This ability made him indispensable to Ed and Louise Hillary. And their work for his people gave them a very special place in his heart. Men like Mingma, wives like Angduli, became integral parts of this new life of Louise Hillary, just as she was of theirs.

One close friend summed up: 'Louise complemented Ed so very well. She was very extrovert, very socially adept and sure of herself. She was the yeasty element that made the bread rise.' She had done more than marry a man, she had also married a job, a job which grew bigger every year, and also a role in a distant community.

The Hillary family orchestra — Peter, Belinda and Sarah try out instruments their father had brought from Nepal.
NZ HERALD

Louise wearing Nepalese jewellery in 1968.
NATIONAL LIBRARY OF NZ

Louise assembles goods for a Kathmandu bazaar organised to raise money for the many Hillary aid projects in Nepal.
NEWS MEDIA

wine
ROLLS

Finally — and it was, sadly, finally — the family left to spend a full year in Nepal, living in Kathmandu while Ed and his team worked their way through their biggest project thus far, rebuilding the airstrip at Phaphlu and then completing a hospital there at Solu.

Before she left, Louise Hillary confided to friends like the Mulgrews as she had done so often before that she hated flying. She had told my wife, Valerie Davies, earlier: 'I'm terrified of being aloft and afloat — I like to be in control of myself and able to cope. I don't feel I am, in a boat or a plane.'

In February 1975, eighteen-year-old Sarah flew back to Auckland to continue her varsity studies. A month later, Peter, twenty, left with a friend, Simon Maclaurin, to travel in Kashmir and Assam. Sixteen-year-

old Belinda was attending a Nepalese girls' school and doing New Zealand Correspondence School lessons.

Like so many phases of their life, times were busy for Ed and Louise Hillary. In two years, he had built dozens of schools and the Phaphlu hospital project was the second, but far from the last, he would build. Their friends and family were always around them. Jim and Phyl Rose, Louise's parents, were in Nepal. Brother Rex Hillary was a key figure in the hospital project.

Ed Hillary was still smiling about a dinner he had shared with Lord Mountbatten and Prince Charles, part of their visit to the coronation of the King of Nepal and how Mountbatten had talked of his belief in UFOs, how he had actually seen one from the bridge of one of his warship commands. Years later, when Ed Hillary told me of this conversation, he chuckled again as he recalled how Prince Charles had given him a half-wink as his uncle's eloquence on the subject reached new heights.

Peter Mulgrew's wife June and their daughter Robyn were in Nepal. One day late in March, they flew in from the mountains to be met at Kathmandu airport by Louise and Belinda Hillary before the Mulgrews left for Auckland. There was so much to talk about. The Hillarys planned a walking tour a few weeks later. There were last-minute messages for daughter Sarah back in Auckland as Louise and Belinda waved their good friends off.

A few days later, Louise and Belinda Hillary took off in the same Pilatus single-engine aircraft that had flown the Mulgrews into Kathmandu. It crashed a minute or two later, killing them, the New Zealand pilot Peter Shand, and two other passengers. Ed Hillary had his first inkling of disaster when a helicopter, rather than the overdue Pilatus, landed with the news at Phaphlu airstrip. Shattered, he walked to where Jim and Phyl Rose were also awaiting the arrival. It was unbelievable, but a fact which had to be faced. They all flew back to the crash scene in the helicopter, 'to the last resting place of the two people I had loved most', as he would later write of it.

He told me later: 'I was numb with sorrow all the way. Louise and I sometimes talked about death but it was me that was meant to die — not Louise.'

Peter Hillary heard the news in Jorhat in Assam and flew back to Kathmandu. His mother and sister Belinda had been cremated beside the river the night they died, 31 March 1975.

June and Robyn Mulgrew learnt of the disaster when they reached Auckland. Sarah Hillary, at the family beach house at Anawhata on

Sir Edmund and Lady Hillary catch up with their diaries at the Seli monastery.
EVENING POST

Debris of the plane crash that killed Louise and Belinda Hillary.
ASSOCIATED PRESS

Sir Edmund at the crash scene.
ASSOCIATED PRESS

How explorers and adventurers spend much of their time — Edmund Hillary dealing with a mass of paperwork that goes with the territory.

NZ WOMAN'S WEEKLY

Greeting son Peter, Rob Hall and Gary Ball after their Everest climb in 1990.

NZ WOMAN'S WEEKLY

The face on the banknote.
NZ WOMAN'S WEEKLY

Auckland's rugged west coast, heard the news first from June and Peter Mulgrew. Mulgrew left for Nepal immediately with Sarah, later returning with the ashes. They were scattered along the Anawhata coast that had always meant so much to Louise and her family.

Peter Mulgrew had written of his apprehension as he watched the Pilatus aircraft take off with Louise and June those years before. For him, premonition had become horrifying fact, and there was more to come.

For Rex Hillary, the crash had double impact. He had known Louise for so long, since before Ed's marriage, and he was required to identify her and Belinda, his niece, at the accident scene. Like all those around him, he was shattered by their deaths. Each mourned in their own way — Angduli, the ever-grateful wife of Hillary's dearest Sherpa Mingma, made a hillside shrine for Louise and goes there to burn incense and to pray. Rex decided his role was to finish the hospital they were working on. Like others close to Ed Hillary, he feared that the terrible impact of Ed's loss might drive him away from the life he and Louise had shared.

Even the faithful Mingma had doubts. When Hillary — in deep shock and mourning — left the area, now dotted with altars to the irreplacable friend the Sherpas had lost, Mingma expressed fears that his beloved leader, like Louise, would never return. He had good reason. Ed Hillary was devastated. People who had seen him in the good times, like Graeme Dingle, saw him as 'changed for ever'.

Hillary later talked in an interview of those times, how he felt that nothing could ever ease the pain. 'People would say to me that time was a great healer, but I simply did not believe it. In fact, time did ease the grief somewhat. I realised that there was no use bemoaning the fact or blaming God or the pilot or anything else. They were dead, what would have been the point in that? That has been my attitude since.

'I just look back on those wonderful years that I had with them, how fortunate I was in having that period of my life when there was that tremendous amount of love and affection. A lot of people never have that opportunity. I did, and I appreciate that. The sadness that was a burden, I just had to put up with.'

He did, and those near him, who had known him a long time in situations calling for courage and strength of mind, saw in him in those times reserves of strength even they found surprising. Often in his life, he and his family had faced the fact of death, but always in relation to his lifestyle. There had always been the potential for disaster. Now, it had come and he had been both survivor and loser. With Peter and Sarah — now an art conservator at the Auckland City Gallery — he went to Europe and on to the United States before returning to the emptiness and the memories of Auckland, to the coastal house where they had all shared so much and where Sarah had first heard the news.

And then, five months after the tragedy, Hillary headed back to the Himalayas. He realised that the only way of countering his grief and frequent bouts of depression was to throw himself back into the work that had been so much part of their shared life.

By December, Rex and the other workers had finished the Phaphlu hospital with its medical wards, its solar heating system and its power generation unit. Ed Hillary was quietly delighted, as pleased as it was then possible for him to be about anything. As Graeme Dingle summed up: 'He just seemed permanently sad.' But even those close to Ed Hillary did not realise how sad.

Hillary himself talks of walking alone in the high places he loved where no one could see his tears, of wondering if life was worth living. His children had to learn to cope, too, and that was not easy. Peter Hillary has described feeling 'demolished . . . stunned . . . directionless'. He has spoken of realising that he and his father now had to establish a new relationship.

Ed, who had talked so openly to me and to others, and has written too, about the obviously stormy life he had with his father, is just as direct in his accounts of his father-son relationship with Peter. He talks and writes about 'the usual teenage problems . . . self-centred and bloody-minded independence . . . on the few occasions I thumped Peter he certainly didn't like it but I can't say it made any difference to his behaviour. I was the one who ended up feeling miserable.' He seems to frequently use the description 'pain in the neck', although by a mid-1993 *Listener* interview, he conceded that parents can also fit that definition.

Those comments all rather bear out that description of one of his closest friends, Peter Mulgrew: 'He [Ed] can appear surprisingly thoughtless, even careless of the feelings of others.'

Clearly life as the son of a world-famous and very high-profile father was not easy for Peter. The boy who learnt to climb first in a Mt Eden quarry not far from his father's old school, Auckland Grammar, went instead to King's College to avoid the 'Sir Edmund's son' label — as if a change of school really meant that! Well, that was the theory.

It obviously didn't work. When he took school holiday work, he used a false name. Ed clearly wanted Peter to have the university education he had dropped out of before the war. Peter, just as plainly, wanted to be himself — with a life of challenge. Thousands of fathers will identify with their battle over motorbikes too. The generation gap was as marked in the Hillary house as anywhere else.

Peter's inevitable compulsion to climb was sometimes a link and at other times an apparent negative factor. Ed would be quoted later as saying he felt 'complete indifference' about Peter's decision to take up mountaineering. Yet, at times, the father himself seemed the catalyst. One of Peter's earliest climbing memories, even before that first climb with his parents in the Himalayas two years later involved Ed and the seemingly ever-present Mingma.

The Sherpa was in New Zealand on a holiday Ed had set up for him. The three of them attempted a snow peak called Mount Fog across Lake Wanaka in the South Island, the ten-year-old roped between the two Himalayan veterans. Finally, they turned back, but for a 'completely indifferent' father that seemed a rather pointed childhood initiation.

By the time Peter Hillary dropped out early from an Auckland University geology degree course, there seemed little doubt where he was going. He had already climbed extensively on both North and South Island peaks, was an inveterate skier as well as a motorcyclist, and would later fly. Peter Hillary was, like his father at that age, an action man.

His climbing took on a new thrust in the years after his mother's Himalayan air crash. Among the most noteworthy was an ascent of 12,239-foot Mt Cook's East Face with Australian climber Fred From. It was the beginning of a deep friendship and a significant climbing partnership, which was to end in death — on Mt Everest.

It was, in the mountaineering and positive sense, all uphill from there, as the young Hillary tried tougher peaks, more difficult routes, higher altitudes. He made a 5000-mile Himalayan traverse over 10 months with Graeme Dingle. He even took his new wife, Australian drama teacher Ann, to Everest on a honeymoon expedition.

With the danger came difficulties and disasters. Five team members died on Peter Hillary expeditions to the Himalayas between 1979 and 1984. With Fred From and Adrian Burgess, he was beaten back only 700 feet from the summit of Lhotse in a brilliant attempt without oxygen that took them to 27,100 feet. Attempting the west face of Amadablam, a party he led was swept by an avalanche. His friend Ken Hyslop died and Peter suffered a broken arm and other injuries.

Three years later, Peter Hillary and Fred From survived another disaster on Makalu when two of their party, Dunedin climber Bill Denz and Australian Mark Moorehead, died in separate accidents before the assault was abandoned. Then, less than a year later, Fred From and another Australian, Craig Nottle, died in two falls while attempting Everest in a party headed by Peter Hillary. Fred From fell while searching for Nottle, who died on his twenty-third birthday. It had been a year of death and despair for Peter Hillary.

Ed Hillary's reaction was: 'I always had the feeling Peter was constantly attempting things and routes that were extremely difficult. As a consequence, he was often unsuccessful. He has always been impatient and can be almost aggressive in his attitude. On Amadablam, it was a hazardous route but they almost made it and that's what keeps them going.'

Peter Hillary and his wife Ann — they went to the Everest region on their honeymoon!
NEWS MEDIA

It did keep Peter Hillary going — finally all the way to the top of Everest, which he scaled with fellow New Zealanders Rob Hall and Gary Ball of Christchurch on 10 May 1990. Peter was the 279th climber to follow his father to the summit, and one of fifteen that day. In May 1992, thirty-two reached the top on one day. In May 1993, three New Zealanders, — Rob Hall again, his wife Jan Arnold and Warkworth farmer John Gluckman — were among a record thirty-seven to make the top, including a three-climber, all-woman team from Nepal. It was the same week that a British ex-commando, Harry Taylor, made the first solo climb without oxygen.

But none of those climbs can match the combined Hillary feat — having reached the top that day in 1990, Peter Hillary called his father on a satellite radio link, first man ever on the peak talking to his son now also on the peak. That was real history.

It was obviously a fairly typical Hillary phone conversation. Peter had said after some disappointment at his father's rather muted reaction to his major climb on Mt Cook: 'He is well known among his friends as one not overly disposed to spiels of superlatives, so his reaction seemed more an interested acknowledgement to me in my state of near euphoria.' Ed's reaction after Peter's Everest climb sounded much the same. It was, he said, the longest chat he'd had with his son in years!

Peter's memory is hazy about what exactly he said to his father. 'We just chatted about the climb and the relative difficulties and the conditions . . . I was pretty impressed with the Hillary Step, the crux of the climb [this was the final rock wall Ed Hillary had talked about so graphically]. A lot of these communications using satellite technology need to be interpreted somewhat. You are dealing with people who are feeling pretty relaxed, probably have a mug of coffee in their hand, at one end of the conversation and people at the other end who are pushed to their absolute extreme. That was pretty much our situation.'

Which sounds as if this conversation was rather short on superlatives, too.

Everest men Peter and Edmund Hillary with Sherpa Tenzing together for a 1982 television documentary.
NEWS MEDIA

A great moment of shared pride as Peter Hillary announces his father's inclusion in the New Zealand Sports Hall of Fame in December 1990.
EVENING POST

Shortly after this, one magazine writer gave this as her interpretation of that father-son relationship: 'Peter talks about his father with pride and affection. But Sir Edmund, who clearly does not find it easy being a parent, admits that he and his son are not close. As Peter struggled to build his own identity, his father withdrew and the gulf between them remained. Sir Edmund was pleased at Peter's success on Everest "because it was obviously important to him . . . Peter is very independent in his views and his outlook as I was at that age. I learned very early on that gratuitous advice was not accepted with great enthusiasm. He is very much his own person. He has made his own decisions and chosen his own challenges and way of life."

'Sir Edmund says Peter is in his prime for high-altitude climbing [he was then thirty-five] "but personally I'd like to see him spend a little more time at home involved with his family and normal activities".'

It's an interesting reaction from a man who, in his mid and late thirties was spending many months at a time away from his family in either the Himalayas or Antarctica. Alter the date and it could have been a rather crusty Percy Hillary talking about his son Ed — except that he would also have added 'the bees' to those needing his son's involvement. It is also the voice of a man who has always had trouble expressing his emotions, a difficulty exacerbated by the death of his beloved Louise.

It was the reaction, too, of a man who had accepted that his high-altitude days were long over. When, as some sort of climber emeritus, he went with an American party that attempted but failed to scale the East Face of Everest from Tibet in 1981, all the old and worrying symptoms recurred at the 17,000-foot base camp and he was forced to withdraw down the mountain. He had double vision, chest pressure, hallucinations and had difficulty speaking. (A number of the same party returned two years later to achieve the summit by this route.)

Father-son relationships are often difficult for outsiders to understand. All the more difficult when the older man is famed and his talented son must inevitably live in the shadow of that fame. Even more difficult when the son seeks and gains fulfillment in the specialist field that his father has dominated. Add in a great shared grief.

Perhaps, after all, the measure of their real, unspoken and at times unrecognised relationship is caught in the photograph of their two faces in the moments when Peter was called on to nominate his father to New Zealand's Sporting Hall of Fame in 1990. Or their reunion years before when Peter returned, injured and shattered by failure, from the high slopes of Amadablam. 'Dad was there. We looked at one another. There seemed nothing to say, just a feeling of love and concern.'

CHAPTER THIRTEEN

Partners

Just as like attracts like, Ed Hillary drew to him, and to his life, extraordinary men — men like Hunt, Lowe, Fuchs and Mulgrew. More than simply men — he also married two outstanding women.

To follow the much-loved and legendary Louise as a second Lady Hillary would have been beyond most women. Yet, when Ed Hillary married June, the widow of his great friend and fellow adventurer Peter Mulgrew, in 1990, it was a match as perfect as the first. Where, at an earlier stage of his life, Louise had provided stimulus and new horizons, and had been part of new adventures, June Mulgrew, Lady Hillary, was a calm and serene personality at a time when the pattern of his life had changed. They had shared much, separately and together, most notably grief.

From the late 1950s, the Hillarys and the Mulgrews had been constant friends. The husbands shared their high-profile adventures, while Louise and June had the common link of young families and absent husbands. Robyn Mulgrew remembers the Hillary family as somehow always there — in good times and in bad. And the bad were very bad.

There was first the terrible physical and emotional adjustment forced on June and Peter Mulgrew, and daughters Robyn and Susan, by the crippling events on Makalu in 1961. June Mulgrew flew to Kathmandu to find Peter with his legs putrefying from gangrene, his hands badly frostbitten, dependent on regular pethidine and morphine to somehow mute the pain. She became his nurse, handling the non-stop injection routine, trying to lift the spirits of an exceptionally active man now facing a life of

A reunion after the horrors of Makalu — Peter Mulgrew and June with Edmund and Louise Hillary.
ROBYN MULGREW

The Hillarys and the Mulgrews share a joke with the Governor-General, Sir Arthur Porritt, at a 1969 garden party.
NEWS MEDIA

34

severe handicap and frustration. Later, she helped him accept the trauma of amputation, and supported him in his successful battle with the pethidine addiction.

There were years of struggle as he adjusted first to his artificial legs and then to the effect they had on his life. Peter Mulgrew was a man of tremendous courage, but he also needed and got the support of an equally strong and courageous woman. She was always there, to buoy up his spirits, to literally pick him up when he fell, as he did often in the first days of his new limbs, to be his typist when — almost as a matter of therapy — he told his story in his book *No Place For Men*. Certainly, there was a place in his life for a woman like June.

Between them, they painted and decorated a new house. When, a few years later, he took up yachting, the women of the family crewed for him. He had fingers amputated and others never uncurled from the freezing effects of that Makalu ice and snow, and he had problems with rigging. Robyn's memory was: 'When we were sailing, someone had to pull the anchor up — he couldn't do it.'

He loved sailing and was a champion, always happiest in his boat. When Tenzing came to New Zealand, one of his greatest memories was a sail with Mulgrew in his One Tonner *Young Nick*. Tin legs and all, Peter Mulgrew took a crew around Cape Horn and the South American coast, trying but failing to convince Himalayan mates Ed Hillary and Mike Gill that they should sail with him. But Makalu party companion Wally Romanes did go — another of the people who could not resist challenge.

The death of Louise Hillary brought the two families even closer together, if that was possible. Peter Mulgrew was perhaps the only friend to whom Ed Hillary could reveal the true extent of his grief during those terrible years. And June Mulgrew was always there for both of them.

Sometimes, courage was not enough for Peter Mulgrew. When he attempted to pick up something of the old life, joining a Hillary climbing and building expedition to Nepal, the truth became too obvious to him. He could never go back, he told his family. He could not bear to be helped and even carried as he now had to be. His old climbing spirit rebelled against the hard facts of his new life. Memories of climbs from the Matterhorn to Makalu were all that remained.

There were other changes. He left the navy and began a new executive life in private enterprise electronics. The marriage broke up. Then came Erebus. Peter Mulgrew was one of a panel of expert commentators who flew on Air New Zealand scenic flights to the Antarctic — Hillary had flown on one earlier. In November 1979, Peter Mulgrew was one of the 257 passengers and crew who died when an Air New Zealand DC10 crashed on the slopes of Mt Erebus.

Once more, the Mulgrew and Hillary families were caught up in grief

Lady Hillary with friend Dorje.

NEWS MEDIA

and support. Robyn Mulgrew describes it as a massive shock. 'You never really recover, you just learn to cope.' June Mulgrew did that, as Ed Hillary had.

The last years had not been easy for her or for Peter Mulgrew. His personality had changed. He was difficult at times, living with worsening problems with his legs, growing increasingly fearful that he would finish up in a wheelchair. His daughter Robyn says, 'This seemed more likely as time went on, and I don't think he would have let that happen. We always presumed that one day he would go down and get on his boat, go away and we would never see him again. That's what we all thought.

'For him to die as he did, on that Erebus flight, seemed right in three ways — he went on a mountain, in the Antarctic and he went very quickly. It was so much better than it would have ended up if he had become disabled.

'He never thought he would be on a mountain again, or in the Antarctic. It was perfect for him.'

It was from this background of shared family lives, disaster and grief, that Ed and June's deep friendship moved on to a partnership and finally marriage. Certainly, it was no surprise. In a typically distinctive decision, Hillary had taken June, by then his 'constant companion' as the media called the relationship, to New Delhi as his social secretary when he was appointed New Zealand High Commissioner there in 1984.

It was an arrangement which would have rocked precedent and protocol with anyone other than Hillary. As it was, bureaucrats, diplomats and the public accepted it as thoroughly logical. As always and for good reason, Ed Hillary, his life and decisions, remained beyond criticism.

The new marriage, the new life, which Ed Hillary, in his typically under-stated way, describes as giving him 'a great deal of pleasure and companionship', brought a carryover from the best of the past.

When members of the 1953 Everest expedition met once again in Kathmandu in 1993, forty years after the triumph of their team and those moments in the lives of Hillary and Tenzing, eighty-two-year-old Lord Hunt hailed the life work that Hillary had given to the Sherpa people in the Solu Khumbu region as 'the justification of our ascent has become the work of Hillary'.

A young Sherpa university graduate, who owed his education to the work of Hillary's Himalayan Trust, said of his mentor: 'You rose above personal interest and displayed your profound humanity by your selfless service to the people of Solu Khumbu.' As he did, more leaders of Sherpa

June Mulgrew, Lady Hillary, wearing her newly invested Queen's Service Medal, May 1990.
DOMINION

communities were in the queue seeking more aid from that trust. Not far away was the new monastery of Thyangboche, replacing the historic building destroyed by fire four years before. It, too, was a tribute to the work of Hillary. He travelled the world raising more than $600,000 to help fund the rebuilding.

We talked about those trips, and others before and since, late one afternoon at his home, as June Mulgrew ironed a quota of large shirts ready for him to fly out to the United States that night. His work for the Sherpa people was — typically, he didn't use the expression — a labour of love. 'A very close affection and relationship with the Sherpa people,' was how he described it. 'I enjoy them enormously. I love the area and naturally I want to help them, as you would any friends.' He talked then as he had before, and has since, about the deep sense of responsibility he feels towards the Sherpas and the drive he had to continue his fund-raising and other work for them. 'I think I'll probably go on doing this until I can't walk, because of this deep sense of obligation.'

Others support him. Some have done for many years, like his brother Rex. As Ed Hillary left Nepal after that fortieth anniversary visit, Rex was arriving again, heading to Phaphlu hospital for one of his regular maintenance checks. This time, it was the solar heating which needed his attention. Rex is particularly proud of that project, built in a year. Like Ed, he talks of a great sense of achievement. 'I don't know how I did it.' And he rattles off detail of the wards and the laboratories, the accommodation for families, all directed with Hillary drive and expertise and involving more than 10,000 hours of unpaid work by the local people.

The only previous major hitch was one Rex quickly detected on an earlier trip. The toilet system was blocked because the Sherpas were using pebbles instead of toilet paper and dropping them into the lavatory bowls!

Repairing Khunde hospital after earthquake damage, building bridges, new classrooms for Khumjung school, one of nearly thirty schools set up, a water system here, a drainage system there, have all been Hillary projects over the years. The Himalayan Trust spends around $500,000 a year, most of it raised solely because of Ed Hillary. No wonder that the Sherpa people love and honour him almost like a god.

New Zealanders don't raise people to such status anywhere near as easily. The last demi-god was Michael Joseph Savage, more than fifty years ago. But, if they stop short of deification, New Zealanders have only praise and the highest level of public respect for Ed Hillary. Only three things prevented him being Governor-General. He was neither a woman nor a Maori when either or both were a prime requisite for the post, and he would probably have declined anyway. He is a man who has always had difficulty hiding boredom and, as he says, tends to counter the risk of it by heading off into yet another unknown.

With monks amongst the fire-wrecked Thyangboche monastery. Hillary would fly the world raising money to rebuild it.
EVENING POST

A junior teacher reads a petition from sixty children asking for a school in their area.
NEWS MEDIA

Above: Rex Hillary, the veteran of years of Himalayan aid projects, wears around his neck a piece of rock from the first climb of Everest, specially mounted for him by a craftsman in Kathmandu.
NEWS MEDIA

Right: Sir Edmund checking progress on the airstrip at Khunde hospital with Dr Paul Silvester, a New Zealander working at the hospital through Volunteer Service Abroad.
THE PRESS

Left: The roof goes on yet another Sherpa schoolhouse.
THE PRESS

Following pages: At work on the Thami schoolhouse near Ama Dablam.
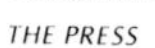
THE PRESS

Another major factor in Hillary's being overlooked for the highest honour is that politicians who make decisions in these areas have always found problems in handling, much less making the best use of, Ed Hillary. He does not fit the established patterns. He does not comply with set rules. He has viewpoints and expresses them. His very public disagreements with two National Party Prime Ministers showed that.

Ed Hillary told a group of Auckland secondary school head prefects in June 1967 that there was too much dishonesty in national and international politics. He urged them to 'bring a little more honest-to-God morality into politics and Government at all levels, nationally and internationally. It horrifies me the way a Head of State can one moment deny vehemently that his country is carrying on some particular action, and then a couple of days later and with complete calmness admit the whole thing.' He talked about 'dishonesty of utterance in Government'.

Years later, and still puzzled at the vehemence of the response, he told me he actually had his mind mostly on the double-dealing involved in United States policy in Vietnam and surrounding states. But his text at the time certainly had Wellington in his sights as well: 'We seem to be experiencing expediency and sheer dishonesty every day. We get it to a lesser level in this country. It happens far too often and gives a bad example to all of us in our lives.' He said his generation tended to accentuate the gap between rich and poor countries.

Prime Minister Keith Holyoake reacted angrily and with typical pomposity, demanding that Hillary substantiate or withdraw. Hillary refused and a few days later got a decidedly public cold shoulder from Holyoake when both men appeared at a meeting, of all things, of Volunteer Service Abroad in Wellington.

The fact was — and Holyoake knew it — that, as in so much of his life, Hillary held the high ground. The theme of New Zealand failure to help bridge the gap between rich and poor nations was one he came back to again and again.

Two years later, he took up a Holyoake statistic about New Zealand spending 0.25 per cent of its gross national product on overseas aid. It was, Hillary said, the lowest of all affluent nations and simply not good enough. He told a Rotary conference in Hamilton: 'Sometimes people say to me why don't Asians and Pacific Islanders or someone else, wherever they may be, do something for themselves? Why don't they pull their fingers out and get cracking like we do? And then that person pours himself another whisky and water and tells you how hard life was when he was young.'

Giving the product a personal recommendation — Ed Hillary tests a new bridge in Nepal.
EVENING POST

Tense moments when the Prime Minister, Keith Holyoake, snubbed Sir Edmund at a VSA meeting in Wellington after his criticism of political morality.
NEWS MEDIA

Sir Robert Muldoon, as Prime Minister, once tried to wave Hillary's criticisms away, saying he knew as much about national finances as Muldoon knew about mountain climbing. That sort of cheap political shot had no hope of stopping a man with Hillary's reputation for pressing on. He just went on pricking the national conscience.

He called for the country to be led from '. . . the last twenty years which have been a period of political, moral and spiritual doldrums . . . we suffer from a lack of political leadership at the highest level . . . I don't really blame the many idealistic young people who have become cynical and depressed about our society.'

Then, in a criticism of racial prejudice in all its forms, Hillary threw his support behind the strong public pressure to break rugby ties with South Africa, which was, at that time, admitting Maori All Blacks as 'honorary whites'. This was strong stuff in rugby-crazy New Zealand.

In this outspoken and very public mood of criticism, he also added his name to those supporting Labour leader Bill Rowling — Citizens for Rowling, the group was named — in stormy political infighting in the mid-1970s. Yet Edmund Hillary was, as his father before him had been, very much his own man.

He had no illusions about politicians. He had, he said once, been thumped on the back by a succession of Prime Ministers and told that if there was anything he wanted, simply to ask. 'I finally did ask for some money for the Sherpa hospitals. I didn't get the money and I haven't been thumped on the back since.' Not only not patted on the back. Having been snubbed by Holyoake he was also ignored publicly by Muldoon who refused to shake his hand, turned his back and walked away after yet another of the Hillary criticisms. Possibly this one: 'If New Zealanders want a story to bring tears to their eyes, they should try suggesting to the Government that it should spend more money on helping people overseas. You will learn more about external balances of trade, decreased prices, the financial pinch and responsibility to taxpayers than you thought possible.'

He was so much a man on the outer with politicians that, when the then Labour Prime Minister David Lange rang him one day in 1984, Ed Hillary asked him a second time who it was who was calling. The Lange call was to offer him the post of New Zealand High Commissioner at the New Delhi posting the Lange Government re-established. He accepted, for, next to Nepal, India was very special to him. The Ganges trip had charmed him. It was an enlightened appointment, but for Hillary, at sixty-five, and for the country, it was at least ten years too late. The nation should have been exploiting the energy and charisma of its foremost citizen of the world long before.

Returning to India took him back that much closer to those peaks and people he clearly loved so much and had given so freely of his time to.

Even then, there was a slight delay. Hillary needed a day or two before taking up the post in April 1985 to meet Neil Armstrong, first man on the Moon, in a memorable moment at the North Pole! Just another name, another historic place, in the life of an adventurer: the first man on the summit of Everest, first to make a motorised journey to the South Pole, meeting the first Moon explorer now at the North Pole. From that first moment on the high peak, he had always seemed larger than life, and this was yet another occasion in a life without equal in his generation.

Meanwhile, then and later, he remained a rallying point for the millions wanting the protection of the environment. Through organisations like the World Wildlife Fund, he represented to ordinary men and women a universally accepted role model of a concerned citizen of the world.

Through his Himalayan Trust, he gave those needy people of the high slopes a better life, their health was improved and protected, their education advanced, their homes enhanced. Their environment was given new life through the extensive tree-planting project that the trust has embarked on, and through Hillary's own concern, which is sometimes close to anger, over the despoiling of the Everest region through the rubbish that thousands of trampers and climbers leave every season.

In New Zealand, he was equally outspoken on the unemployment issue, the right to work, health changes, and the sale of New Zealand assets. 'I don't by any means have all the answers to these problems, but I do believe that we human beings have a right and a duty to speak out, to express our concerns on these matters.'

In a revealing insight into the real Hillary, he put aside his snub from that rather peppery Sir Robert Muldoon those years before and attended his funeral in 1992, admitting that he wept a little in the process. He told an interviewer later: 'I cry easily. I wasn't one of Sir Robert's greatest supporters by any means, but there was something about him I had to admire all the same.'

No one would doubt that Ed Hillary would cry easily. In a life of triumph, he had suffered great loss too. In a very public life, he had few close friends. While his many great qualities had been revealed by the pattern of his life, he had also been exposed by it as well, seldom out of the public eye even in the Himalayas where a Hillary journey so often became a triumphal journey through admirers — and those who sought something of him.

His few closest friends, of course, know him best and their assessments are typically matter-of-fact and genuine. None more than George Lowe: 'A great New Zealander. It's as simple as that. An ordinary man who is at the same time a most extra-ordinary man.'

New Zealand's High Commissioner to India with Umu Sanker, son of the high commission's gardener, and Babu Gomez, a watchman.

NZ HERALD

Left: Old friends, old memories — Edmund Hillary and Tenzing. Having climbed Everest, they went to the top of Wellington's Mt Victoria together when the Sherpa visited in 1971.

EVENING POST

Right: Trying on an old, familiar hat before it was handed to Canterbury Museum in 1972.

THE PRESS

Below: Edmund Hillary with friendly and fierce masks from Nepal.

NZ HERALD

CHAPTER FOURTEEN

At the Top

Forty years on, nothing is the same. Even Everest was being trimmed — by a few feet from its 29,002 feet given by the Royal Geographical Society and the generally accepted figure in 1953, to under 29,000 in an Indian survey shortly after the first climb. Then, it lost another few feet according to a Chinese remeasure in 1974. In these days of metric measurement, Everest is now accepted as 8846.1 metres above sea level, putting it back above the 29,000-foot mark by some twenty-four feet.

They are fractional and unimportant movements. At the same time, the peak has, according to an opinion that must be respected, lost something important. Lord Hunt said after the swarming assaults of May 1993 that Everest was losing its mystique, 'the whole mystery of it has been diminished'. This was a sad commentary, but understandable. The Everest climb, although still very difficult, has become almost a group travel event. Aided by new techniques, super-light-weight equipment and well-blazed routes, nearly five hundred have reached the peak since Hillary and Tenzing.

Hunt's reaction came in a week when Everest also revealed something of its old power. Lobsang Tenzing, nephew of the great Tenzing Norgay, reached the peak, along with thirty or more from various expeditions on that day, before beginning a descent with the British solo climber Harry Taylor. Taylor reached the lower slopes safely. Lobsang, who had become separated from him, died in a fall about 200 metres above the South Col, which his uncle knew so well. His was the 117th death on Everest.

On the same day, Tenzing Norgay's grandson Tashi had to abandon his attempt only 400 metres from the summit when he lost his goggles and was temporarily blinded. It was as if Everest — forty years after Hillary and

Tenzing — was showing that, while it had been climbed, it had never really been conquered. And names meant nothing to it.

One name, though, remains unchallenged in its greatness. One reputation is higher than ever before. One feat is still acclaimed as without equal. Edmund Hillary still stands on a personal peak of worldwide and unqualified respect.

He is a man whose integrity and courage have never been questioned, whose loyalty to friends is legendary, whose determination is unmatched, whose strength both of mind and of body at his peak made him unique among his fellows. Hillary is a man held in awe by his peers, a giant among giants.

Like anyone, he has characteristics which sometimes catch you by surprise. Understandably, he can be off-hand and even dismissive, particularly when his amazingly trying pattern of travel and obligations coincides with an inquiry he feels is fatuous, a contact which is pointless or trivial. By contrast, he can also be understanding and warm in the most unlikely circumstances.

He is a heavy critic, often in print. He does not hesitate to speak and write his mind. His first and largely unamended reactions to Vivian Fuchs, his later assessment of the Fuchs leadership at crucial stages of the polar crossing, are typical. So is his blunt recall of the worst periods of his relationship with his children first given in his early writing, repeated in his joint book with Peter, *Two Generations*, and still a feature of even the fortieth anniversary interviews with him.

But he also appears to accept what little criticisms he has encountered with forbearance. Certainly, at no stage of the Pole dash controversy did he reach for the headlines to justify himself. He bore the post-Everest adulation of Tenzing with public stoicism, and fiery British media criticism of his sprint to the Pole with studied ignore. For all that, he is a man of undeniable singlemindedness, which at least two close observers during my research translated as ruthlessness.

Achievement came relatively late to him, at thirty-three, but he quickly adjusted to the experience and has gone on achieving as no one of his generation and few in any modern era have. From all this has come an aura of greatness about him that never diminishes.

It was already there when I first met him after Everest, gangly, loose-limbed, friendly and seemingly always smiling. It was with him in the dark years of his sorrow. It still remains. Interestingly, I was most aware of it on probably the lowest peak Hillary ever climbed. It was thirty-seven years after we had first met.

Following pages: Forty years on — Hillary the new hero in 1953, the elder statesman of the 1990s.

C. HALLIDAY & AUCKLAND MUSEUM/*DOMINION*

We were on the summit of Mt Hobson, one of Auckland's dormant volcanic peaks only a few hundred metres from his Remuera home, photographing him for the *Living Treasures* series. He was co-operative but rather abstracted. Back at the house, June was packing for that journey to raise money for Thyangboche he was to leave on in a few hours. With the shots taken, he excused himself and headed home. There are moments in your life that you remember for some special, sometimes unexplainable factor. This was one of them.

I can remember others. There was a day in the 1950s when as parliamentary reporter of the *Auckland Star* I watched an old Bob Semple, one of the last survivors of the great first Labour governments, walk down the steps of Parliament Buildings into retirement with a miner's helmet, reminder of his union past, under his arm. There were the moments in Covent Garden when Fonteyn made her first entrance in *Cinderella*. And, a few days earlier, I had stood on a London pavement and watched through a sports store door as Sir Jack Hobbs, then nearly eighty, helped a schoolboy choose a cricket bat from his shop's stock. There is a physical reaction to such moments which is indescribable but always there, a swelling of emotion, a tingling both of skin and awareness.

That was how it was on Mt Hobson that day as I watched that bulky figure in rough trousers and a zipped jacket, which would have been at home at Camp Seven, walk off down the path, a carved walking stick in his hand. He strode purposefully like someone with something on their mind, a task to do, a journey to make. It was unmistakeably Ed Hillary. He carried with him an indefinable presence, a sense of history.

In that glimpse, before he disappeared over the low hill, he was so many things: once a lean man in a striped, home-made sunhat peering up at a great peak, then 'knight of the realm' as an old Percy Hillary had referred to him, then a builder with a blueprint of another schoolhouse in his hand, once a man determinedly on his way to the South Pole, come what may, now the face on New Zealand's five-dollar note.

Or simply, as George Lowe described him, Mr Everest.